PACIFIC COAST SUBTIDAL MARINE INVERTEBRATES

A Fishwatcher's Guide

by
Daniel W. Gotshall
and
Laurence L. Laurent

SEA CHALLENGERS • MONTEREY, CALIFORNIA

1979

A SEA CHALLENGERS PUBLICATION

FRONT COVER

Orange sun star, *Solaster stimpsoni*, Photo By Daniel W. Gotshall.
Printed by Dai Nippon Printing Co., Ltd., Tokyo, Japan.

ISBN: 0-93011802-2 (Paperbound)

Library of Congress Catalog Card Number: 79-64128
First Edition

SEA CHALLENGERS
4 Somerset Rise
Monterey, California 93940

Phototypesetting and pre-press production by Padre Productions.

FOREWORD

Pacific Coast Subtidal Marine Invertebrates is a book we have long needed in the diving world of the west coast. It will be useful both to the beginner and to the old-timer, no matter how experienced. For who of us (except perhaps for the authors) is so knowledgeable as to have seen all of the undersea environments from Alaska to Baja California? In a sense, we are all neophytes when we enter the water in a new place for the first time. We may see many familiar plants and animals, but new environments are always replete with delightful surprises and unknowns. This field guide will be of great assistance for many of us, helping to clarify these mysteries of the ocean.

Dan Gotshall and Bud Laurent very wisely decided not to describe even a major portion of the hundreds of invertebrates that may populate the sea floor in a given region. A compendium of all these species would probably be comprehensible only to a handful of specialists. The authors have here presented 161 of the commonest invertebrates. Each invertebrate is clearly illustrated with a color photograph showing the animal in nature. Such photos can be extremely helpful because appearance in the native habitat is often very unlike an artist's rendition of a preserved specimen. *Pacific Coast Subtidal Marine Invetebrates* also provides a short description of every animal, outlining characteristics such as color, shape, anatomical details, typical habitats, and depth and geographical ranges, where such information can be used for identification. Gotshall and Laurent have done an excellent job in choosing between what to present and which organisms to exclude. In most cases, readers will find in this book a majority of the conspicuous and interesting invertebrates encountered in the first few visits to new diving sites. If your find is not in this book, you will likely have collected a specimen whose identification is tricky and requires some skill and perhaps concentrated effort, coming through a specialized treatise.

It is probably necessary, but a bit regrettable, that in books such as *Pacific Coast Subtidal Marine Invertebrates* we encounter the professional competence of the authors but receive almost no exposure to their personalities. Dan and Bud are wonderful guys to dive with and to recount sea stories afterwards across a steaming pizza and glasses of flavorful wine. The breadth of their activities is reflected somewhat by my earliest experiences with them. I first dove with Laurent on an idyllic coral reef in Tahiti, while my introduction to Gotshall occurred in the middle of a ferocious, bone-chilling gale which he casually described as "about average here in Point Arena".

Both are unassuming individuals. One hardly realizes they are in the room—that is, until someone cracks a joke—Gotshall has an explosive laugh that rattle the windows.

Gotshall and Laurent have carved successful careers as diving marine biologists with the California Department of Fish and Game. They are widely recognized for their competence as biologists and for their firsthand knowledge of the underwater. *Pacific Coast Subtidal Marine Invertebrates* is certain to add significantly to this recognition. My own understanding of the sea has been enriched by my long friendships with Dan and Bud. I am most privileged that they asked me to prepare this foreword to this book.

Wheeler J. North
Corona del Mar
April 20, 1979

DEDICATION

This book is dedicated to all the divers over the years who have shared with me the continually new experience of the last frontier—the underwater world. But, most of all, it is dedicated to my wife, Ann, for her patience, understanding and love of all living things.

Daniel W. Gotshall

I dedicate my portion of this book to my wife, Sandra, for her continuing patience, aid and support and to our lovely children, Adam and Simone, my sources of constant delight and occasional perplexity.

I would also like to dedicate this to friends of my Moss Landing Marine Lab years where we learned, mostly from each other, the skills of becoming biologist-divers: Shane Anderson, Rich Ajeska and Dave Lewis. Memories of the dives we made and survived, with leaky and primitive gear, remain far more vivid than many of the more recent and more comfortable dives.

Laurence L. Laurent

TABLE OF CONTENTS

INTRODUCTION

The diver who descends into the Pacific coastal waters for the first time is immediately confronted with a panorama of plants and animals which are both fascinating and confusing. Those of us who live on the Pacific coast of North America are privileged to have on our continent's doorstep one of the richest assemblages of marine life in the world. Tropical regions may offer the attractions of warm water, mind-boggling visibility and schools of brightly colored fish, but these regions can't begin to compete with the diversity, color and sheer lushness of plants and animals found on the rocky bottoms of the nutrient-rich waters of our coast. Our fervent hope and intent is that this field guide will help to dispel some of the confusion regarding the diverse invertebrate kingdom, as well as assist divers and others in achieving the satisfaction of learning fresh significances of animal behavior and finding species new to their experience. Under such circumstances, we are all scientists regardless of our professional callings, trying to understand the order of our natural world. It is to augment this natural scientific inclination that we have written this guide to the marine invertebrates—the animals without backbones.

There are several thousand invertebrate species along this coast, many of which can only be identified by specialists. The 161 species included in this guide were selected using the following criteria: 1) the animal must live at least partially within the depths to which divers are limited; 2) the animal must be common in a fairly wide range between Alaska and Baja California; and 3) the animal must be easily visible and recognizable to a diver. In making this selection, we were forced to exclude many common species, especially among the smaller crustaceans and molluscs and also difficult-to-identify organisms such as sponges, hydroids, bryozoans and tunicates. For these animals we direct the interested diver to more advanced publications, many of which we have listed in the reference section. In addition, there are many excellent identification guides readily available which cover snails, clams, and chitons.

This guide properly should only be considered as an introduction to the subtidal invertebrates of the Pacific coast. Our purpose is to enable the sport diver and others to identify most of the larger invertebrates encountered in the area covered by the guide. It is designed also to enable the biologist diver quickly to learn the more easily recognized animals and/or to confirm an identification made from a key such as *Light's Manual*. Finally, for each animal, we have the most recent geographic and depth ranges available. This type of information on invertebrates has remained hidden in biologists' field notes and scientific literature for far too long; our hope is that this guide will serve as a base from which more accurate ranges can be established.

Information of this type is of vital importance to biologists studying marine communities. For example, if an area under study for a sewer outfall lacked a particular species, even though the area provided the proper habitat and was well within the geographic and depth limits of the species, the biologist's task, if possible, would be to determine if the sewer outfall was responsible for the absence or disappearnce of the animal. Because of the importance of this geographic and depth range information, we would sincerely appreciate learning of any corrections or additions to the data presented in this guide about which any of our readers may know. All such validated information will be incorported in future editions and credit given to the person providing the information. The best method for verifying a specimen is to collect it (assuming it is legal to do so) and to have it examined by an authority from a university or museum.

Since one of the purposes of this book is to aid marine enthusiasts in putting names to organisms, it is appropriate to discuss some of the problems of nomenclature. As you will discover, we list each animal by both common and scientific names. Common names are included because many people prefer them, finding scientific names unwieldly and hard to remember. There are problems with common names, however, one being that they are often regional, varying from place to place. For example the *Cryptochiton stelleri* (# 66) is referred to variously as the giant chiton, the Pacific chiton and the gumboot chiton. In limiting ourselves to a single common name, we have chosen the one most frequently used or the name which best describes the animal. Another problem is that for many invertebrates there simply is no common name. In these cases, we have taken the liberty of arbitrarily assigning a common name for those who prefer to use them.

For clarity and positive identification, the best way of referring to an animal is to use its scientific name, but this, too, presents problems. The science of nomenclature, called taxonomy, is a practice based on a scholarly understanding of anatomical form and function, plus a thorough grasp of taxonomic literature. It is also a science in flux because, unfortunately, its ideal of absolute precision has not often been achieved. Many factors contribute to this, among them being legitimate disagreements about proper placement of species, inaccessibility of pertinent literature and even, on occasion, previous errors by taxonomists that have remained in the literature. But all of that notwithstanding, the important thing is that the scientific name is universally understood.

Briefly, a scientific name is composed of two parts: the first name (always capitalized) is the genus; the second name (never capitalized) is called the trivial. The genus is equivalent to a human surname, such as Jones or Smith. The trivial (also called "species") is equivalent to a first name, indicating which Jones or which Smith. Together they form what is known as the specific name of an organism.

In an attempt to make scientific names more understandable and meaningful we have also included their translations. This has had mixed results. Using Latin and Greek texts we have done our best and while many of the names yielded readily to an easy and smooth translation, others gave us no little difficulty. If we have made any grievous mistakes in this effort which are discovered by any Greek or Latin scholars among you, we would be grateful for your help in correcting such errors.

How To Use This Guide. If the animal is unfamiliar to you, refer first to the pictorial key on page 14. This key is based on shape, presence or absence of legs and other obvious anatomical characteristics. The animals that fall into a particular category are listed with appropriate page numbers. Drawings are provided to help you in distinguishing particular characteristics. Once you have decided upon a group of animals to which your particular critter belongs, then turn to the page indicated and look at the photographs. In cases where your choice is narrowed down to two or three species, you must carefully consider the identification and range information. Characters are provided that best separate closely identical species and which should help you in deciding which species you observed or collected. The geographic range may also aid you here. For example, if you are diving in British Columbia and your choice is down to two animals, one of which has not been recorded north of Point Conception, California, you can be fairly sure that your animal is the one whose range includes British Columbia. Following national policy, we have used the metric system to designate sizes of animals. However, because all depth gauges are still calibrated in feet, we have continued to use feet for depth measurements.

New Information. We are very interested in learning of your observations where extensions in the geographic range or a new depth or maximum size of a species is noted. Should you observe an animal in an area not listed in this guide or perhaps come across one that sets a new size record, we are most eager to hear about it. Likewise, we would appreciate having our attention called to any published records that inadvertently we may have missed. Please send all such comments and information to the authors at: California Department of Fish and Game, Operations Research Branch, Sacramento, California.

Daniel W. Gotshall
Laurence L. Laurent

Acknowledgements

We could not have completed this book without the assistance of a large number of people. We would especially like to thank the following: Dave Behrens, biologist, Pacific Gas and Electric Co., for his valuable editorial comments on portions of the manuscript and for his confirmation of the identification of many of the nudibranchs; Barbara Bowman, sponge taxonomy biologist, California Academy of Sciences, San Francisco for her identification of some of the sponges, Dustin Chivers, Curator of Invertebrate Zoology, C.A.S., San Francisco, for providing information on many of the invertebrates; John De Martini, Professor, Humboldt State University, Arcata, for the final technical editing; Daphne Dunn, anemone taxonomy biologist, C.A.S., San Francisco, for the identification of the anemones; Ray Emerson, Kinnetic's Laboratory, Santa Cruz, for the notes he provided on *Diopatra*; Bob Given, Director, and Jack Engle, Biologist, U.S.C. Marine Laboratory, Santa Catalina Island, provided unpublished range and depth data for several species; Ann Gothsall, for her helpful suggestions, much of the typing, proofing the galleys, the Glossary, and for her aid and involvement with the literature search; Wheeler North, Professor, California Institute of Technology, Pasadena, for the geographic and depth range data he provided and for confirming or providing identifications for many of the animals; and Carl Sisskind, Scripps Institute of Oceanography, for the final copy editing. Contributions of corrections for this second printing were received from Kirk Stoddard, Robert Sellers, Joan Stewart, Leslie Harris, William Austin, and Craig Fusaro.

Many others assisted us in various ways in collecting data and providing underwater photographs. These include the staff of the Bamfield Marine Station on Vancouver Island, Canada; Jerry Mc Crary and Bill Donaldson of the Alaska Department of Fish and Game, Kodiak; and the staff of the University of Southern California Marine Laboratory on Santa Catalina Island. To all these people, and to any others that we may have inadvertently missed, we offer our humble and sincere thanks for helping to bring this project to completion.

PHOTO CREDITS

Photographer	*Species Number*
Dave Behrens	84, 96
Richard Burge	74
H. Richard Carlson	47, 59, 87
Tony Chess	68
Earl Ebert	75
Robert Given	160
Bernie Hanby	61, 107
Laurence Laurent:	5, 6, 14, 15, 24, 49, 56, 65, 67, 69, 78, 80, 81, 82, 85, 88, 89, 90, 103, 106, 114, 116, 123, 128, 129, 140, 142, 146, 149, 151, 158, 159

The drawings of the maps and animals were rendered by Laurence Laurent.
All other photographs by Daniel Gotshall

GLOSSARY

branchial (brangk' ial) — Relating to the gills, organs for breathing the air contained in water

branchial plume — The featherlike formation of gills in certain marine animals, such as the nudibranchs.

bryozoans — Literally "moss-animals". A group of minute invertebrates, mostly marine, which propagate by budding and which form branching, encrusting or gelatinous colonies of many small polyps, each having a circular or horseshoe-shaped ridge bearing ciliated tentacles.

calcareous — Of, like, or containing calcium carbonate, calcium or lime; chalk-like in substance.

carapace — The upper shell or bone-like covering of a crab. A shield, test or shell covering some or all of the dorsal part of an animal.

carpal segment — The fifth segment of the leg of certain crustaceans

cerata — The small projections (or papillae) on the back of nudibranchs.

chelae (kē lē) — The pincer-like claws with which some of the limbs are terminated in certain crustaceans as the crab, lobster, etc.

chitin (kīt' en) — A hard compound, the chief constituent of the external coverings of crustaceans and insects.

chromatophore — A pigment cell capable of contraction and expansion with consequent change of color as in the octopuses and squid.

cilia — Hairlike outgrowths of certain cells, capable of vibratory movement.

colony — A group of animals or plants of the same kind living or growing together in close association and functioning as *one* integrated organism.

commensal — An animal or plant living with another for support or sometimes for mutual advantage, but not as a parasite.

decapod — Any crustacean of the order *Decapoda*, having five pairs of walking legs, including the crabs, lobsters, crayfish, prawns, shrimps, etc.

diatoms — Any of numerous microscopic, unicellular, marine or fresh-water algae having siliceous cell walls.

dorsal — Of, pertaining to, or situated on or near the back of an animal or any organ; as a dorsal fin; the dorsal aspect of the hand.

dorsum — The back; also the back of any part of organ.

Eolid (Aeolid) — A type of nudibranch that possesses numerous fleshy dorsal processes called cerata and lack plume-like gills around anus.

epipodium — In zoology, a muscular lobe developed from the lateral (side) and upper surfaces of the foot of some molluscs.

excurrent siphon — The projecting tubular part of some animals through which water is expelled from the body.

exoskeleton — The hardened, external supporting structure of an animal, usually composed of a horny, bony, or chitinous substance in the form of plates or scales, as in crustaceans.

filter-feeder — An animal that obtains food by straining out minute planktonic organisms as they pass through specialized structures either on or within the animal.

hermaphrodite — An animal having the sex organs of both the male and the female, so that reproduction can take place without the union of two individuals.

holdfast — A rootlike appendage of a seaweed.

incurrent siphon — The projecting tubular part of some animals through which water is drawn into the body.

intertidal — The shore region that is above the low tide mark and below the high tide mark, hence—between the tides.

mantle — The membranous flap of or folds of the body wall of a mollusc containing glands that secrete a shell-forming fluid.

medusa, medusae (pl.) — (1) An organism with a disc or bell-shaped, gelatinous form with tentacles around the edges. One of the stages in the life cycle of a hydrozoan (i.e., a type of colonial animal). (2) A jellyfish.

nudibranch (nōō′ dah brangk) — A suborder of molluscs lacking both a shell and true gills and having external, often branched respiratory appendages on the back and sides.

operculum — In many gastropods, the horny plate which closes the opening of the shell when the animal is retracted inside.

osculum, oscula (pl.) — Exhalent opening of sponges.

papillated — Any small nipple-like projection or process on the surface.

pedicellaria — An external organ resembling a forceps, found in starfishes and sea urchins.

peduncle — A slender, stalk-like part.

periostracum — The external, chitin-like covering of the shell of certain molluscs, which protects the limy portion from acids.

plankton — The billions of microscopic animal and plant life found floating or drifting in the ocean or in bodies of fresh water, used as food by various marine animals.

plumose — Feathery or plumelike.

polyp — A sedentary type of animal form characterized by a more or less fixed base, columnar body and free end with mouth and tentacles. Any of a number of small, flower-like water animals having a mouth fringed with many small, slender tentacles at the top of a tube-like body, as the sea anemone, hydra, etc.

radials	Pertaining to structures that radiate from a central point, as the arms of a starfish.
radula	A chitinous band in the mouth of most molluscs, set with numerous, minute, horny teeth and drawn backward and forward over the odontophore (structure supporting the radula) in the process of breaking up food.
rhinophores	In molluscs, one of a pair of fleshy process of nudibranchs.
rostrum	A beak-like part or appendage.
serrated	Notched on the edge like a saw.
sessile	Refers to the fact that an organism is permanently attached to the substrate in one location.
spicule	A small, hard, sharp-pointed, needle-like piece or process, especially of bony or calcareous materials, as in the skeleton of the sponge.
spongin	A horny substance that forms the skeletons of certain sponges.
stipe	Stalk. A stalk or slender support, a stem-like part.
substrate	In biology, the medium upon which an organism grows. The base or material upon which an organism lives.
subtidal	The area of the coast extending from the lowest low tide to the deepest portion of the ocean.
test	The outside hard covering of certain animals, as the shell of *Mollusca* or the calcareous shell of sea urchins or the thick leathery outer tunic of the sea squirts.
tubercle	In the echinoderms, a small eminence of the body wall which is immediately connected with the spines. In appearance, a small rounded projection or excrescence.
tunicate	Having a tunic or mantle, any member of the family *Tunicata*.
umbilicus	In conchology, a circular depression in the base of the lower whorl or body of many snails.
ventral	Of, near, on, or toward the belly or the side of the body where the belly is located.
zooid	Any one of the recognizably distinct individuals or elements of a compound or colonial animal, whether detached, detachable, or not.
zooid chamber	The minute cavity in the compound or colonial organism where the zooid lives.

METRIC EQUIVALENTS

1 inch	=	2.54 centimeters	1 millimeter	=	0.04 inches
1 foot	=	30.48 centimeters	1 centimeter	=	0.39 inches
1 yard	=	0.91 meters	1 meter	=	39.37 inches (3.28 feet)
(1 fathom [6 feet])	=	1.82 meters	1 kilometer	=	3281 feet (0.62 miles)
1 mile	=	1.609 kilometers			

Artifical Key To Some Pacific Coast Subtidal Marine Invertebrates

1.a. animal free-living....12
b. animal attached....2

(1b) 2.a. soft to the touch....7
b. with hard parts....3

(2b) 3.a. attached by 'stalk', with branches, or attached by base and lacy in appearance....5
b. attached by base, short....4

(3b) 4.a. short, radiating tentacles around a central mouth with a calcareous base
—the solitary corals, p. 34
b. generally with a wrinkled, conical test surrounding beak-like plates; legs, when extended, frilled
—barnacles, p. 44

(3a) 5.a. fan-like, much-divided, medium-hard to the touch
—the gorgonians, p. 36
b. divided, tree-like or shrub-like, quite hard to the touch....6

(5b) 6.a. surface even with regular star-shaped openings, branch tips rounded, color generally shade of red or purple
—the hydrocorals, p. 28
b. either erect and much divided with tube-like openings at the surface, or lacy with window-like openings
—the bryozoans, p. 76

(2) 7.a. encrusting or rounded and clump-like, generally not attached by stalks or columns 8
b. attached by column, tube or stalk 9

(7a) 8.a. smooth, slick-feeling to the touch without obvious body openings
—the compound tunicates, p 98
b. slightly rough to the touch, often with volcano-like openings at surface, or with spicules in evidence
—the sponges, p. 22

(7b) 9.a. mostly feathery in appearance.... 10
b. appearance otherwise....11

(9a) 10.a. attached directly and vertically to substrate, by a short stalk; branching is flat
—the hydroids, p. 26
b. branching of feathery processes usually protruding from a tube
—the plume worms, p. 40

(9b) 11.a. with tentacles (numbering from 10 to hundreds) ringing a central mouth, attached by a generally thick soft column
—the anemones, p. 28
b. attached by long stalk, or at base, with two obvious body openings at apex
—the solitary tunicates, p. 98
c. protruding from sand bottom, usually only seen at night
—the sea pens, p. 36

(1a) 12.a. without obvious hard parts....13
b. with obvious hard parts....16

(12a) 13.a. free-swimming, at least part of the time....14
b. bottom dwelling....15

(13a) 14.a. dome-shaped, often with long, unadorned tentacles
—the jellyfish, p. 40
b. with 8 or 10 tentacles with suction cups
—octopus and squid, p. 76

(13b) 15.a. long, cylindrical, with tube feet
—the cucumbers, p. 94
b. moves on a creeping foot, usually hemispherical or elongate
—the nudibranchs, p. 66

(12b) 16.a. moves about on thin, pointed, jointed legs....17
b. moves about on flat, creeping foot 19
c. moves about on tube feet....21

(16a) 17.a. oval, flattened body, usually with large claws, abdomen folded under
—the crabs, p. 44
b. body elongate, muscular, abdomen extends behind....18

(17b) 18.a. usually small, (<20 cm total length) one or two of the walking legs is a much-jointed claw
—the shrimp, p. 44
b. usually large (>20 cm total length), clawless
—the lobster, p. 52

(16b) 19.a. with shell composed of 8 sections
—the chitons, p. 52
19.b. with shell entire, not in sections....20

20.a. shell cap-shaped, sometimes with an opening at the apex
—the limpets, p. 54
b. shell in form of flattened coil, shell opening nearly same diameter as shell itself
—the abalone, p. 56
c. shell a conical or rounded coil; opening usually ½ or less of shell diameter
—the snails, p. 58

(16c) 21.a. with five or more arms....22

b. animal spherical and spiny
—the urchins, p. 78

(21a) 22.a. arms thick and tapering, with or without short spines over entire surface
—the sea stars, p. 82

b. arms thin, fragile, usually only with lateral (side) spines along their length—central portion of animal smooth and circular
—the brittlestars, p. 92

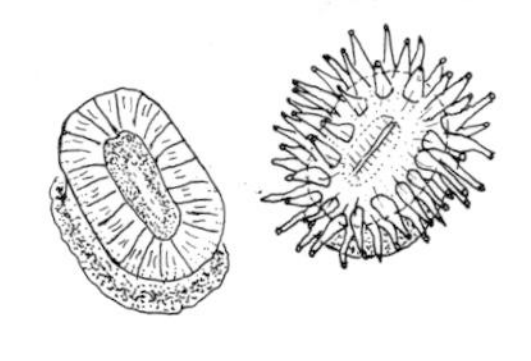

4a SOLITARY CORALS

4b BARNACLES

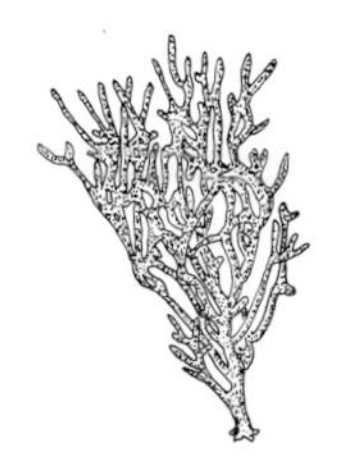

5a GORGONIANS

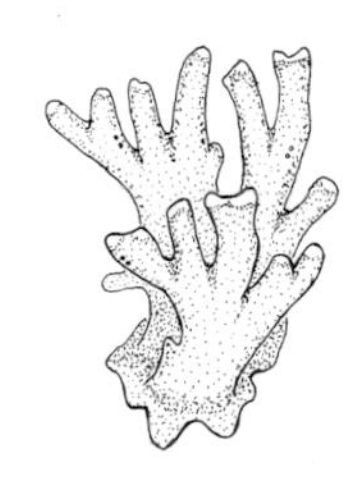

6a HYDROCORALS

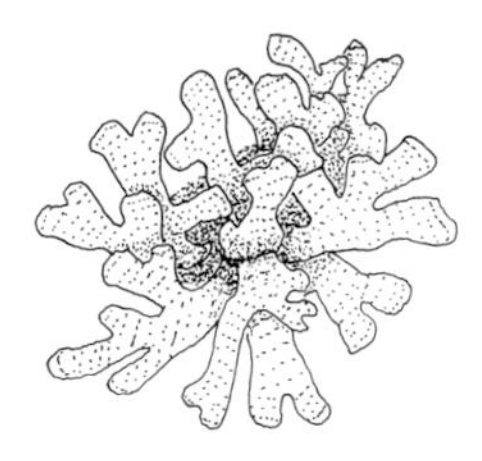

6b BRYOZOANS

6b BRYOZOANS

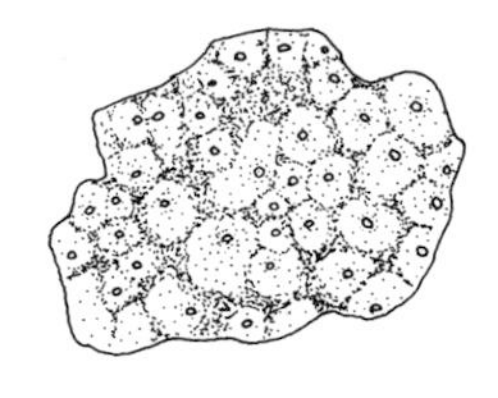

8a COMPOUND TUNICATES

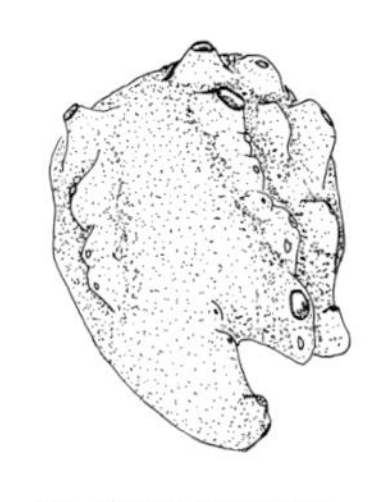

8b SPONGES

8b SPONGES

10a HYDROIDS

10b PLUME WORMS

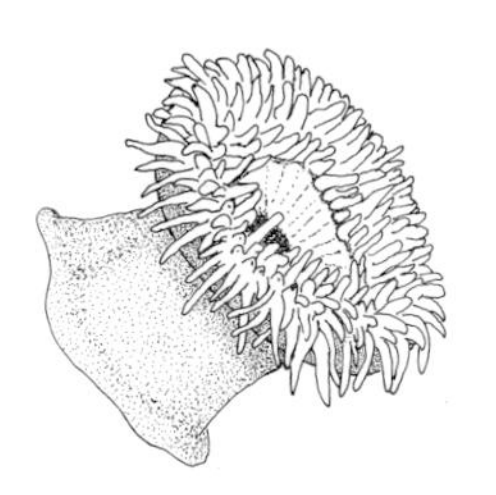

11a ANEMONES

11b SOLITARY TUNICATES

11c SEA PENS

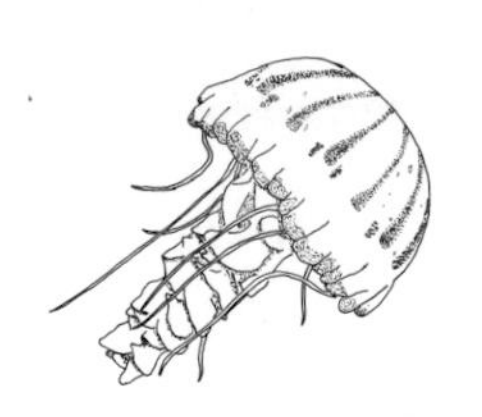

14a JELLYFISH

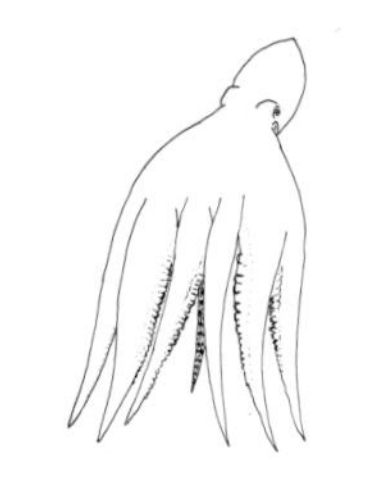

14b OCTOPUS & SQUID

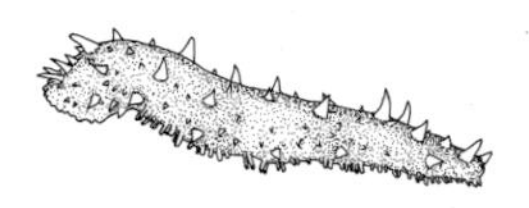

15a CUCUMBERS

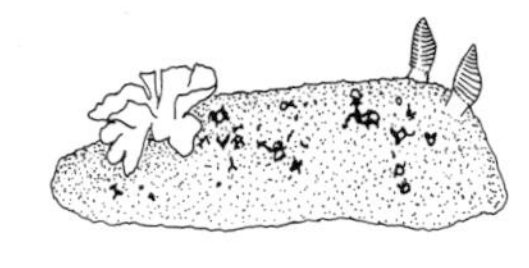

15b NUDIBRANCHS

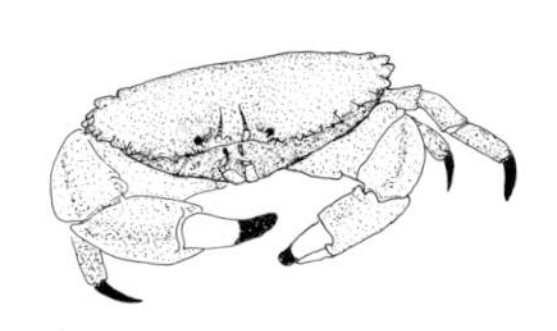

17a CRABS

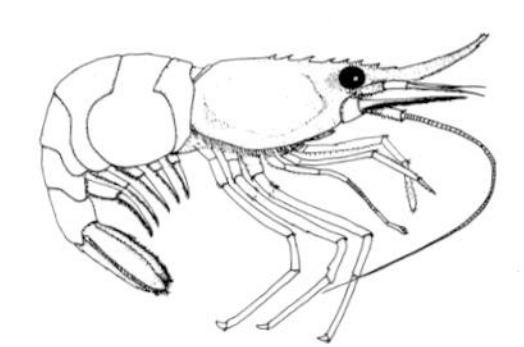

18a SHRIMP

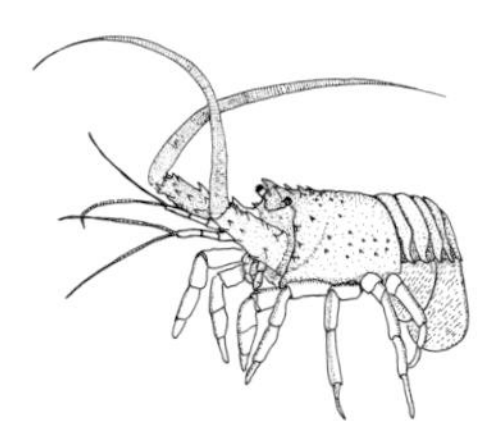

18b LOBSTER

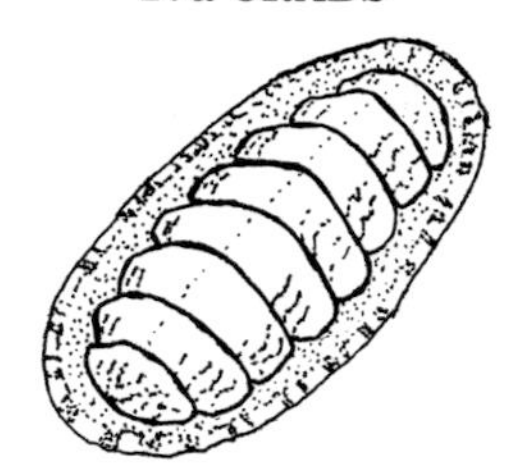

19a CHITONS

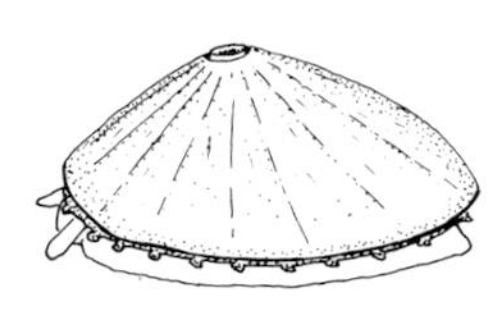

20a LIMPETS

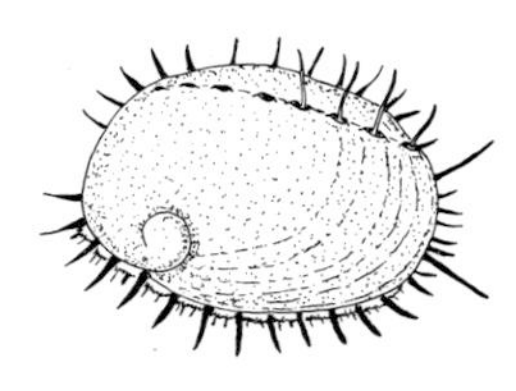

20b ABALONE

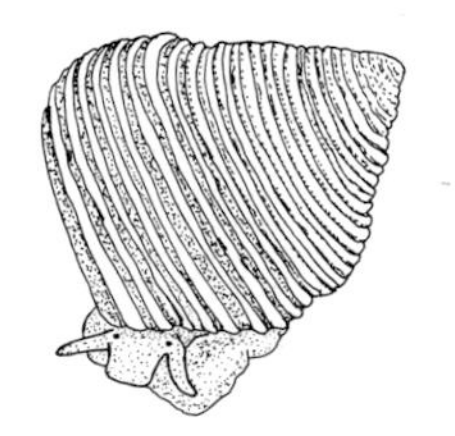

20c SNAILS

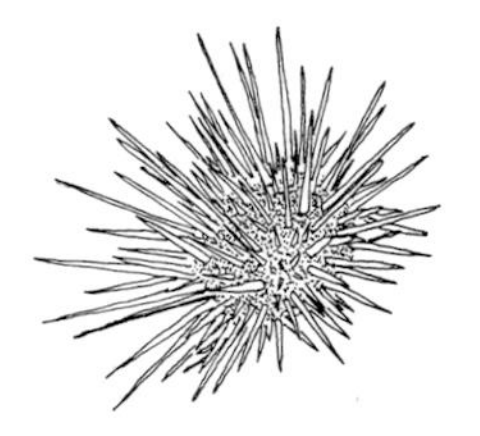

21b URCHINS

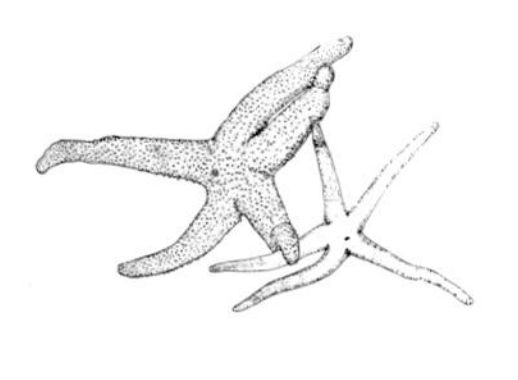

22a SEA STARS

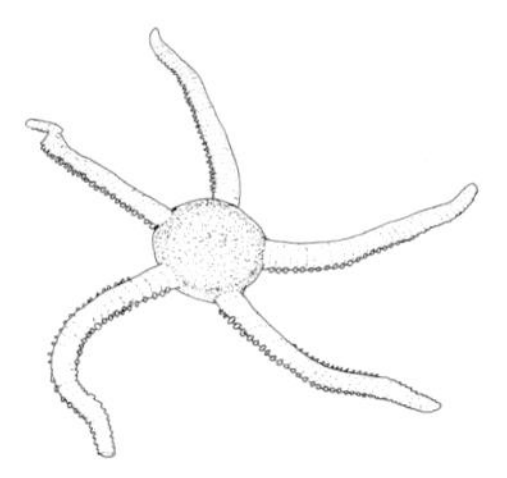

22b BRITTLESTARS

PHYLUM DESCRIPTIONS

PHYLUM PORIFERA

Sponges

Sponges, which come in a multitude of shapes, sizes and colors, constitute one of the more primitive invertebrate groups. All are attached to the substrate. Sponges lack organs, being composed instead of a loose aggregation of cells. Most have a skeleton composed of calcareous or siliceous spicules, spongin fibers or various combinations of these elements. All sponges obtain food by filtering out small plants and animals from water as it passes through numerous small incurrent pores in the sponge.

The filtered water leaves the sponge through a few large apertures called oscula. Sponges, in turn, are preyed upon by nudibranchs and snails. Sometimes encrusting sponges resemble encrusting colonial tunicates (page 98) and a hand lens must be used to determine to which group they properly belong.

Identification of most sponges can be extremely difficult and usually requires use of a microscope. We, therefore, have included here only the large, conspicuous species commonly encountered by divers.

PHYLUM CNIDARIA

Hydroids, Anemones, Jellyfish, Gorgonians, Corals

In this group of animals, we encounter a more complex body structure. The cnidaria possess both a mouth and a digestive cavity and to aid in their feeding, an extension of the body wall has developed into tentacles. Another unique characteristic of this group is that they all possess specialized stinging cells called cnidoblasts. These stinging cells (nematocysts) contain the poison which can be released as protection or to stun prey. The nematocysts are usually found in attached as well as free-swimming forms. Hydroids, corals, gorgonians, anemones and sea pens are attached to the substrate most of their lives. However, during their early development the Hydrozoa and and Anthozoa have free-swimming stages (medusae). The true jellyfish, on the other hand, are free swimming throughout their lives. We have included common subtidal members for all three classes of cnidarians: the Hydrozoa, Anthozoa and Scyphozoa.

PHYLUM ANNELIDA

Polychaete worms

This book includes those worms most likely to be seen by a diver—the tube-dwelling worms. Even when these worms are retracted, their tubes are usually visible.

The tubes of these worms are variously constructed of calcium carbonate, of parchment, of sand grains or of debris. These tubes usually extend into the substrate, or are firmly affixed to the solid surface.

Most tube-dwelling worms are usually restricted to their tubes. However, some worms build tubes and yet can leave them, at least partially, while foraging. One of these is *Diopatra ornata* (#45).

The sedentary worms either filter-feed or deposit-feed which compensates for their lack of mobility. Sabellids possess feathery processes on their heads which greatly increase their food-capturing surface area. These cilia-covered radioles, or "feathers", also serve for respiration.

These worms reproduce by releasing eggs and sperm into the water where fertilization and larval development take place.

PHYLUM ARTHROPODA

Barnacles, Shrimps, Crabs, Lobsters

The crustaceans are characterized by hard, jointed exoskeletons. Their growth involves "molting", whereby the old exoskeleton is discarded for a newer, larger one. The marine crustaceans include barnacles, shrimps, lobsters and crabs, plus several groups not covered here. The barnacles belong to the subclass Cirripedia, order Thoracica, and the shrimps, lobsters and crabs belong to the subclass Malacostraca, order Decapoda (ten legs).

Barnacles, especially the acorn and goosenecks, or stalked forms, feed by filtering particles from the water with their frilled legs. The decapods include herbivores, carnivores, scavengers and suspension feeders.

The barnacle covered in this book is hermaphroditic, and internal fertilization occurs through copulation. Each animal possesses a greatly extensible penis with which to mate with a neighbor. Development of young occurs in a brood chamber of the adult, until they are released to lead a short, free-swimming existence.

Decapods fertilize their eggs also by copulation, which occurs just after molting. The female carries the developing eggs on specialized abdominal appendages (during which time she is called "berried") until the larvae are released. Sometime during the larval period the young hatch and subsequently fend for themselves as they complete development. Most crabs and shrimps pass through several larval stages.

The major difference between shrimps, lobsters and crabs is that the abdomen of shrimps and lobsters is muscular and extended posteriorly, whereas in the crabs it is flexed forward under the anterior part of the body. One notable exception to this difference is in the hermit crabs where the swollen, naked abdomen extends, usually within the protective confines of an empty snail shell.

There are two major groups of crabs: the anomurans, whose fifth pair of walking legs is hidden because they are commonly turned up under the carapace. Their eyes are located between the antennae. The brachyurans (true crabs) have five pairs of walking legs, and their eyes are external or lateral to the second antennae.

PHYLUM MOLLUSCA

Chitons, Limpets, Abalone, Snails, Clams

The molluscs are a very diverse and, from the standpoint of evolution, successful group of animals—they have been on earth for at least 500 million years. Members include bivalves (clams, scallops and mussels), gastropods (snails with and without shells), chitons (resembling gastropods, but with a shell composed of eight plates or valves), and cephalopods (squid and octopus). Scaphopods, or "tooth shells", are also molluscs but are not covered here.

The one external feature uniting this disparate assemblage is the presence of a mantle which commonly secretes the shell and a muscular foot that is generally used for locomotion. In the bivalves the foot is usually hatchet-shaped; in the gastropods and chitons it is flat and adherent and in the cephalopods it is modified into grasping tentacles.

Methods of feeding within the molluscs are as varied as the group. Bivalves are usually filter-feeders and use modified gills to trap and to select appropriately sized food. Chitons are mostly herbivorous. Abalone and limpets are herbivores. The more evolutionarily recent shell snails and nudibranchs are carnivorous. The cephalopods, with some species reaching the largest size of all invertebrates, are strictly carnivorous. The chitons, gastropods and cephalopods all possess a file-like organ at the back of their mouths called a radula which they use for rasping plant and animal tissue. The number and shape of teeth on a radula are much modified, and are often used to aid in identification.

Most bivalves reproduce by a coordinated release of eggs and sperm into sea water; larval development is external to the parent. A few forms like scallops, heart clams and some oysters are hermaphroditic. The chitons fertilize externally, and some brood their young. Abalone and limpets reproduce through external fertilization. Advanced snails possess organs for internal fertilization through copulation.

All opisthobranchs (nudibranchs, sea hares) are hermaphroditic; reproduction involves reciprocal fertilization through copulation with organs located on their right sides. Eggs are deposited in a variety of strings and ribbons, depending on the species.

In cephalopods, the mating process is elaborate and involves a fascinating display of changing skin color and, in the octopus, skin texture as well. The male possesses a specialized arm, called a hectocotylus, which transfers a sperm packet into the mantle cavity of the female. After fertilization, the female deposits the eggs, contained in gelatinous packets, on hard objects on the sea floor. Unlike the squid, a female octopus guards the eggs and maintains water circulation around them, preventing suffocation and/or infection. During this vigil she will not eat and often starves to death.

PHYLUM ECTOPROCTA

Bryozoans

This lesser-known group of animals contains about 4,000 species, making it one of the largest groups. Most bryozoans are colonial and all are attached to a substrate. Individuals making up the colony are called zooids. Zooids lack special organs for respiration, circulation and excretion; instead they are made up of a hard outer shell composed of a chitinous layer overlying a thicker layer of calcium carbonate and an interior cavity, or coelom, containing a U-shaped digestive system. Feeding is accomplished by a ring of tentacles surrounding the mouth. Cilia on the inner surface of the tentacles move food captured by the tentacles to the mouth. Most bryozoan zooids are hermaphoditic; that is, each individual contains both ovaries and testes. Fertilization takes place when eggs and sperm are released into the water or, as in some species within the coelom where the embryos are brooded.

Bryozoans form either flat encrusting sheets or upright colonies that resemble small coral heads or stalked hydroids. Because a hand lens or microscope is required to distinguish the encrusting species, we excluded them. We have included only four easily recognizable, upright species whose colonies form typical formations.

PHYLUM ECHINODERMATA

Sea Stars, Urchins, Cucumbers, Brittle Stars

This group of animals, except for the cucumbers, is characterized by having a skeleton of calcareous plates, or ossicles, on the outside of their bodies. In the cucumbers, this outside skeleton is not apparent as the ossicles are microscopic. Echinoderms usually have tube feet. The feeding behavior of echinoderms ranges from sea cucumbers that extract organic matter from sediment, to plant eating sea molluscs, other echinoderms and even fish. Most echinoderms reproduce by releasing the eggs and sperm into the water where fertilization takes place. Four classes of echinoderms occur commonly in the shallow waters covered by this guide: Sea stars, Class Asteroidea; brittle stars, Class Ophiuroidea; sea urchins and sand dollars, Class Echinoidea; and sea cucumbers, Class Holothuroidea.

PHYLUM CHORDATA

Subphylum Urochordata

Tunicates

There are three groups of tunicates, two living in the water column and one living attached to the bottom. The latter group, the ascidian tunicates or sea squirts, are covered here. The tunicates are placed in the subphylum urochordata of phylum chordata because during some part of their life history they possess three chordate characteristics: a notochord (around which the spinal column forms in the vertebrates), a dorsal nerve cord, and a pharynx (which becomes gills in fish). The body of a tunicate is enclosed within a protective "tunic", composed of a type of cellulose.

There are two types of sea squirts—solitary and colonial. The bodies of solitary and colonial tunicates are generally large and visible to the diver; however, some may require a microscope in order to be seen. The individual, of colonial forms, is called a zooid and many hundred of zooids together make up the organism known as a colony. The anatomy of solitary forms is identifical to that of colonial forms, except that they occur as much larger single individuals.

Tunicates are mostly hermaphroditic and fertilization occurs within the animal. In some species self-fertilization can occur ,but in others the processes of protandry (changing from male to female) or protogyny (changing from female to male) prevents this. The eggs of solitary tunicates are shed into the seawater and larval development is completely external. In colonial and compound tunicates, much of the larval development occurs in a brood chamber within the adult. The larva, during the ascidian's larval stage, is called a "tadpole"; it is free-swimming and has a tail which contains a notochord and nerve tube.

PHYUM PORIFERA

SPONGES

1. **GRAY PUFFBALL SPONGE** ***Tetilla arb***

(little teat tree?)

Identification: This gray-white sponge has a ring of spicules on the dorsal surface.
Natural History: Can be found attached to rocks throughout its range in depths as shallow as tidepools out to at least 80 feet. They are very common in central California.
Size: The average-sized gray puffball sponge would measure 12 by 15 cm. in diameter. Maximum diameter is about 25 cm.
Range: Fort Bragg, California, south to at least San Diego, California.

2. **ORANGE PUFFBALL SPONGE** ***Tethya aurantia***

(Orange *Tethya* [a sea goddess])

Identification: This porous, globose sponge has a color range from orange to yellow which may be hidden by a growth of commensal green algae.
Natural History: Common around rocky reefs, particularly under ledges to depths of at least 175 feet. The purple-ringed top shell (#83) (see photo) feeds on the orange puffball sponge.
Size: Maximum size about 15 to 20 cm. in diameter.
Range: Southeastern Alaska to central Baja California.

3. **FINGER SPONGE** ***Isodictya quatsinoensis***

(Equal-net-like [sponge] of Quatsino [British Columbia])

Identification: This sponge ranges from red to orange in color and has upright finger-like projections, each with a single large opening (osculum) at the tip.
Natural History: Finger sponges can be found around rocky areas in depths of from ten to at least 100 feet. The finger sponge was formally known as *Esperiopsis rigida* by taxonomists, and more than one species may be present in our area.
Size: May reach 20 to 25 cm. in height.
Range: Gulf of Alaska to Point Arena, California.

4. **AGGREGATED VASE SPONGE** ***Polymastia pachymastia***

(Many breasts, thick breasts)

Identification: A single large oscule is apparent at the tip of each white-to-pale yellow sponge.
Natural History: This sponge occurs commonly in vase-shaped groups of 100 or more animals covering an area 60 to 150 cm. in diameter. The aggregations are usually found in areas where there is a light layer of sand over a rocky substrate. Common in Carmel Bay from the very low intertidal out to about 50-foot depths.
Size: May reach 8 cm. in height and 2 cm. in diameter at base of each vase.
Range: Known to occur from Trinidad Bay south to San Nicholas Island.

1. GRAY PUFFBALL SPONGE

2. ORANGE PUFFBALL SPONGE

3. FINGER SPONGE

4. AGGREGATED VASE SPONGE

5. SPINY VASE SPONGE — ***Leucandra heathi***

(Heath's white man)

Identification: Spiny vase sponges have a fringe of spicules around the opening at the tip.
Natural History: This pear-shaped sponge can be observed attached to rocks or other objects from intertidal out to 240 feet, particularly where there are strong currents which supply a steady source of food. The spiny vase sponge belongs to that group of sponges possessing calcareous spicules. Like all sponges, spiny vase sponges reproduce by budding or releasing cells that will form a new sponge.
Size: Size ranges from about 2 to 10 cm. in height.
Range: Point Cabrillo, California, south to Baja California.

6. URN SPONGE — ***Leucilla nuttingi***

(Nutting's little-white-one)

Identification: An urn-shaped, cream-white sponge with a single osculum at the tip.
Natural History: Urn sponges occur in groups of from five to ten or more individuals attached to rocks from the low intertidal out to at least 80 feet.
Size: The average urn sponge is about 4 cm. high.
Range: Trinidad Bay, California, south to Baja California.

7. WHITE FINGER SPONGE — ***Toxadocia*** spp.*

(Bow and lance)

Identification: A white sponge, lacking large oscula that forms masses of finger-like projections.
Natural History: These sponges are common around Point Lobos, California, at depths of 40 feet and deeper.
Size: A single, solitary "finger" may reach 30 cm. in length and 2 cm. in diameter.
Range: The white finger sponge has been observed from central California to southern California.
*More than one species may be involved.

8. RED VOLCANO SPONGE — ***Acarnus erithacus***

(Fleshless robin)

Identification: An encrusting sponge, scarlet in color, with numerous volcano-shaped osculums. The red volcano sponge belongs to that group of sponges that possess spicules either of silica or spongin or both.
Size: Masses grow to about 4 cm. in height and up to 30 cm. in diameter.
Range: Known to occur along the entire California coast.

5. SPINY VASE SPONGE

6. URN SPONGE

7. WHITE FINGER SPONGE

8. RED VOLCANO SPONGE

9. **GRAY MOON SPONGE** ***Spheciospongia confoederata***
(Leagued together wasp-sponge)

Identification: A massive, smooth, gray sponge with numerous moon-crater-like osculums on outer ridge.
Natural History: Common in central California, particularly at depths of from 40 to 60 feet.
Size: This sponge can grow to a thickness of 30 cm. and may reach 450 cm. in length.
Range: Central California south to at least San Martin Island, Baja California.

10. **SULPHUR SPONGE** ***Verongia aurea***
(Golden true [sponge])

Identification: This bright yellow sponge has an irregular surface with a few large osculums.
Natural History: A common sponge of southern California, particularly around the Channel Islands and Santa Catalina Island from the intertidal out to at least 130-foot depths.
Size: Maximum size; 5 cm. thick and 12 to 20 cm. in diameter.
Range: Southern California south to at least Cedros Island, Baja California.

11. **COBALT SPONGE** ***Hymenamphiastra cyanocrypta***
(Hidden, dark blue membrane around star [shaped spicules])

Identification: An encrusting, usually cobalt blue sponge that forms large sheets.
Natural History: This brilliant blue sponge can be observed on sides or rocks, underneath ledges and on the ceilings of caves from the low intertidal out to 190-foot depths. The blue coloration comes from bacteria living on the sponge. When the bacteria are not present, the color is usually orange.
Size: The soft encrusting sheets may reach 90 to 120 cm. in diameter.
Range: Monterey south to Baja California.

PHYLUM CNIDARIA

HYDROIDS, ANEMONES, JELLYFISH, GORGONIANS, CORALS

12. **OSTRICH-PLUME HYDROID** ***Aglaophenia struthionides****
(Splendid ostrich)

Identification: The brown colonies consist of a main axis with many branches that resemble feathers.
Natural History: Hydroids are found wherever ocean currents will insure enough food for survival; ostrich-plume hydroids, always attached, are found on most reefs, generally on the pinnacles. We have not observed them deeper than about 50 feet. Colonies consist of feeding individuals (polyps) equipped with tentacles for capturing food as it drifts into reach, and reproductive polyps which produce a free-swimming medusae. The medusae, in turn, produces eggs or sperm that develop into another colony.
Size: Individual "plumes" may reach 15 cm. in length.
Range: British Columbia to San Diego, California.
*More than one species may be included under the common name.

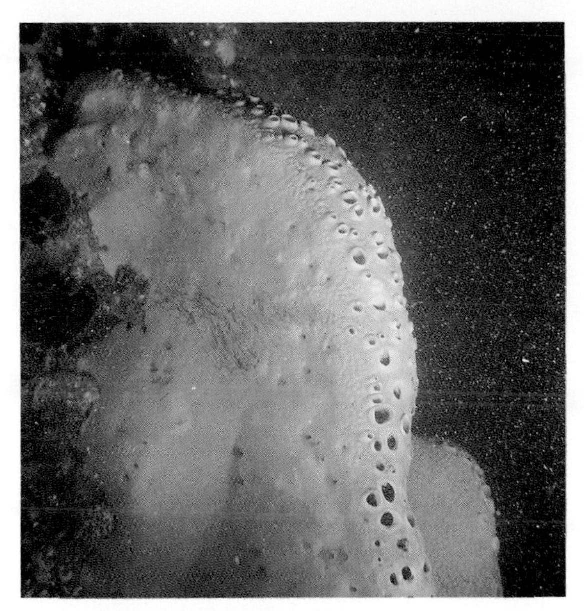

9. GRAY MOON SPONGE

10. SULPHUR SPONGE

11. COBALT SPONGE

12. OSTRICH PLUME HYDROID

13. **BELL MEDUSAE** ***Polyorchis penicillatus***

(Tuft of many testicles)

Identification: This bell-shaped medusa has a single whorl of tentacles on margin. The medusa is transparent with red spots around the rim.

Natural History: Bell medusae are free-swimming members of plankton in our inshore waters. This jellyfish is the asexual product of an unknown species of attached hydroid. Bell jellyfish feed on smaller members of the planktonic community. They are preyed upon by larger members of the plankton and several species of fish.

Size: May reach 15 cm. in length.

Range: British Columbia to central California.

14. **CALIFORNIA HYDROCORAL** ***Allopora californica***

(California's different pore)

Identification: The branching calcareous skeleton resembles staghorn coral and the color ranges from purple to pink-red.

Natural History: Attached colonies of this beautiful hydrozoan are usually restricted to offshore reefs and pinnacles where strong currents provide relatively clean water and abundant planktonic food. The California or blue hydrocoral has been observed as deep as 175 feet. It may take 20 or more years for a colony to reach a height of 30 cm.

Size: Colonies may reach 30 cm. in height.

Range: Vancouver Island south to San Benitos Island, Baja California.

15. **ENCRUSTING HYDROCORAL** ***Stylantheca porphyra***

(Purple column-cup)

Identification: Colonies form in flat encrusting sheets, usually dark blue or purple in color.

Natural History: This hydrocoral occurs from the low intertidal out to depths in excess of 50 feet. Little is known about this encrusting hydrocoral, but growth rate and feeding habits are probably similar to those of the California hydrocoral.

Size: Colonies may reach 60 to 90 cm. in diameter.

Range: Northern California south to about Point Conception.

16. **ROSE ANEMONE** ***Tealia piscivora***

(Teal's fish-eating [anemone])

Identification: The rose anemone has a deep red column with few or no tubercles (bumps). The tentacles are usually white.

Natural History: This large, recently described anemone usually is attached to the sides of rocks, from the low intertidal to depths of at least 100 feet. Rose anemones feed on fish as well as a variety of other animals, limited only by the prey coming within reach of the tentacles. This anemone has been confused with the mottled red-and-green anemone, *Tealia crassicornis* (not illustrated).

Size: May reach 20 cm. in diameter.

Range: Polar seas south to La Jolla, California.

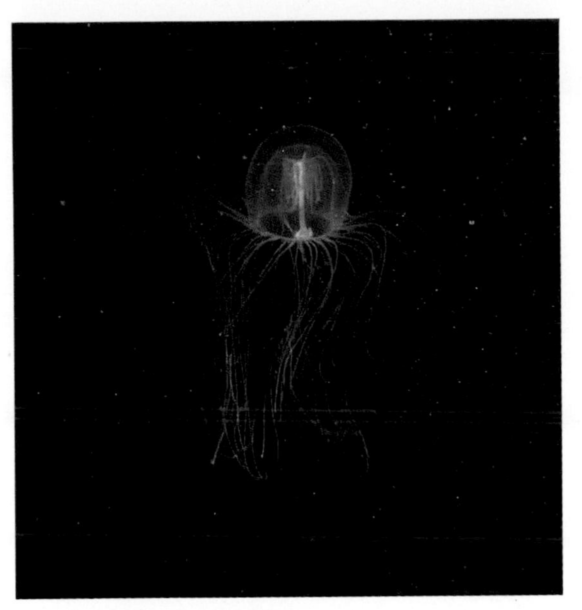

13. BELL MEDUSAE

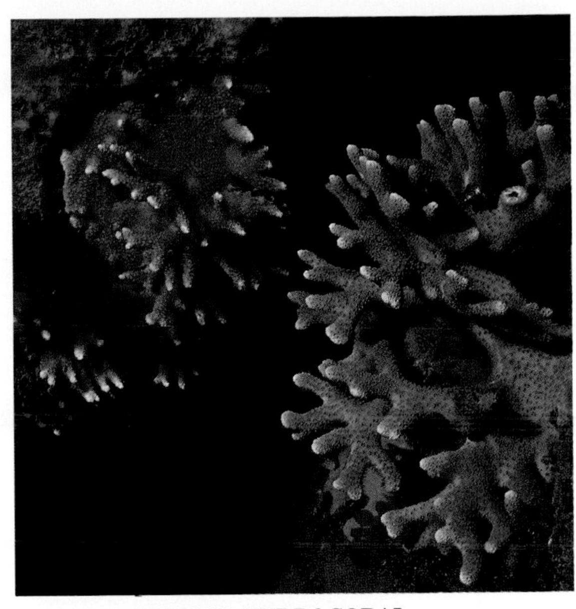

14. CALIFORNIA HYDROCORAL

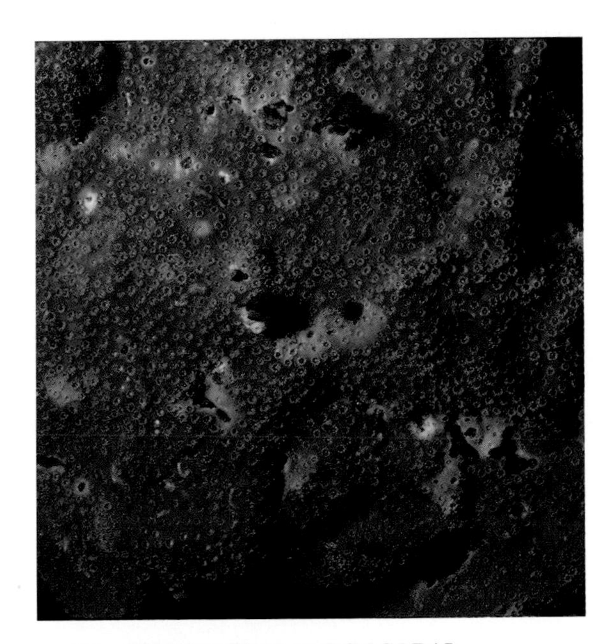

15. ENCRUSTING HYDROCORAL

16. ROSE ANEMONE

17. **WHITE-SPOTTED ROSE ANEMONE** ***Tealia lofotensis***

(Teal's crested [anemone])

Identification: Column a brilliant red with numerous white spots.
Natural History: Found attached to sides of rocks, occasionally in caves and crevices, very low intertidal to about 75 feet. These distinctive anemones feed on whatever animals come within reach of their tentacles, including crabs, jellyfish, and sea stars.
Size: To about 20 cm. in diameter.
Range: Polar seas south to the Channel Islands.

18. **STUBBY ROSE ANEMONE** ***Tealia coriacea***

(Teal's leathery [anemone])

Identification: This anemone has stubby, multi-colored tentacles. The column is densely covered with tubercles which are often covered by shells and gravel.
Size: To about 10 cm. in diameter.
Range: Polar seas to San Diego, California.

19. **SAND-ROSE ANEMONE** ***Tealia columbiana***

(Teal's dove [anemone])

Identification: Possesses long, white tentacles and a column with numerous tubercles covered with attached sand and shell fragments.
Natural History: This anemone is found buried in sand or mud bottoms in depths of 40 to about 150 feet.
Size: Reaches a diameter of 20 to 35 cm. and a height of 25 cm.
Range: Southern British Columbia to Baja California.

20. **GIANT GREEN ANEMONE** ***Anthopleura xanthogrammica***

(Yellow-lined flower)

Identification: A usually solitary uniform green anemone. When found living in crevices away from sunlight, the color is white.
Natural History: Attached to rocks from intertidal to about 100 feet. The green color may be due to the presence of single cell green algae that live within the tissues. Giant green anemones feed on whatever animals come within reach of the tentacles, including fish.
Size: To a diameter of about 25 cm.
Range: Alaska south to Panama.

17. WHITE-SPOTTED ROSE ANEMONE

18. STUBBY ROSE ANEMONE

19. SAND-ROSE ANEMONE

20. GIANT GREEN ANEMONE

21. **MOONGLOW ANEMONE** ***Anthopleura artemesia***

(Diana's lined flower)

Identification: The long slender tentacles range from a pale pink to red. Moonglow anemones have tubercles on upper two-thirds of column only.
Natural History: Attaches to rocks with column buried in sand; intertidal to about 100 feet.
Size: Diameter seldom exceeds 5 cm.
Range: British Columbia to southern California.

22. **PROLIFERATING ANEMONE** ***Epiactis prolifera***

(Fertile on a ray)

Identification: *Epiactis* is primarily some shade of red, but also occurs in green and brown hues and, only occasionally, a lovely lavender. It can be distinguished from another small red anemone, *Corynactis californica*, by the presence of numerous longitudinal grooves in the column and radiating stripes on the disk.
Natural History: Adults are generally found on rocks from the intertidal to about 60 feet along the open coast where there is protection from sand scouring. They can also be found on the less stable habitat of holdfasts of brown kelp and on the blades of red and green algae. *Epiactis* derives its name from its trait of "brooding" its developing young in the column grooves near the base.
Size: Maximum disc diameter about 4 cm. Average diameter nearer 2 cm.
Range: Puget Sound to La Jolla, California.

23. **STRAWBERRY AMENONE** ***Corynactis californica***

(Californian club-ray)

Identification: The only small anemone with club-tipped, usually white tentacles. Color varies from white to pink, orange, red and lavender.
Natural History: This common colonial anemonecan be observed on most rocky reefs and pier pilings where there are strong currents to provide food. Occurs from intertidal to depths of at least 70 feet. These abundant anemones feed on small animals found drifting in the middle and upper part of the water column.
Size: Reaches a maximum diameter of about 2 cm.
Range: Point Arena, northern California, to San Martin Island, Baja California.

24. **YELLOW ANEMONE** ***Epizoanthus scotinus***

(Dark animal-flower)

Identification: These aggregating anemones can be tan, brown or yellow, and have light yellow to white tentacles.
Natural History: Common on offshore rocks where there are strong currents; occurs from low intertidal rocks out to depths of at least 50 feet. Reproduces sexually as well as by budding, hence the usually dense aggregations of individuals.
Size: Diameter of about 2 cm.
Range: British Columbia to central California.
NOTE: There are several other species of zoanthid anemones in southern California, the most common, *Parazoanthus lucificum*, (not illustrated) colonizes on the gorgonian, *Muricea*, possibly leading to the death of the gorgonian.

21. MOONGLOW ANEMONE

22. PROLIFERATING ANEMONE

23. STRAWBERRY ANEMONE

24. YELLOW ANEMONE

25. **WHITE-PLUMED ANEMONE** ***Metridium senile***
(Ancient womb)

Identification: The small, very abundant tentacles form a fine maze. Color is usually white, but yellow, brown and orange are also common, particularly in the northern part of range.
Natural History: Usually attached to rocks and pilings in shallower depths. Occurs from the low intertidal out to depths of 100 feet. Feeds on very small animals that are caught in a maze of tentacles. Reproduces sexually as well as by fragmentation of disc; the fragment then forming a new anemone.
Size: Reaches about 45 cm. in height and 10 cm. in column diameter.
Range: Polar seas to Santa Catalina Island.

26. **TUBE ANEMONE** ***Pachycerianthus fimbriatus***
(Fringed, thick, yellow flower)

Identification: The long, whip-like tentacles and black, parchment-like tube are very distinctive. The color of the tentacles ranges from cream-white to orange to black.
Size: To 30 cm. in height and 5 cm. in diameter.
Range: British Columbia to San Diego, California.
NOTE: More than one species may occur within this range.

27. **ORANGE CUP CORAL** ***Balanophyllia elegans***
(Elegant acorn-leaf)

Identification: This solitary coral has orange tentacles encased in a stony cup of the same color.
Natural History: Attaches to rocks from the low intertidal out to depths of at least 80 feet. These are solitary, true corals.
Size: Diameter of about 2 cm.
Range: British Columbia south to about central Baja California.

28. **BROWN CUP CORAL** ***Paracyathis stearnsi***
(Stearn's cup)

Identification: Look for the almost clear tentacles in a brown stony cup.
Natural History: Occurs on rocks in depths of 30 to 200 feet.
Size: May reach 4 cm. in diameter.
Range: Trinidad Bay, California to Cedros Island, Baja California.

29. **SEA PANSY** ***Renilla kollikeri***
(Kolliker's little kidney)

Identification: The colony consists of a mass of transparent polyps attached to a heart-shaped disc.
Size: Disc reaches 12 cm. in diameter.
Range: Southern California south to at least Baja California.

25. WHITE-PLUMED ANEMONE

26. TUBE ANEMONE

27. ORANGE CUP CORAL
(upper left)

28. BROWN CUP CORAL
(center)

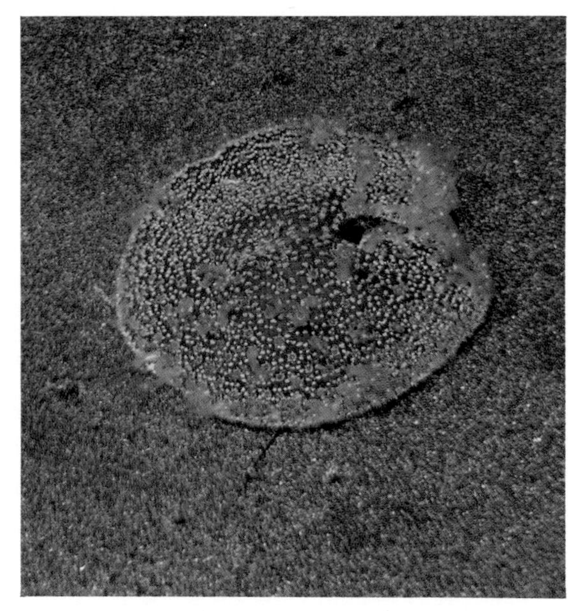

29. SEA PANSY

30. **FLESHY SEA PEN** ***Ptilosarcus gurneyi***

(Gurney's fleshy feather)

Identification: Look for the yellow to orange axis with long, leaf-like lobes and the bulbous buried portion of colony.

Natural History: Found on sand and mud bottoms in depths of about 25 feet (southeastern Alaska) to 225 feet. The feeding polyps of this colonial anemone capture very small animals as they drift near the bottom. The fleshy sea pen produces a strong greenish luminescence when stimulated.

Size: Reaches a height of at least 45 cm.

Range: Gulf of Alaska to southern California.

31. **WHITE SEA PEN** ***Stylatula elongata***

(Elongated little column)

Identification: The white sea pen has a slender, rough to touch, white axis with fluffy, plume-like, white lateral branches.

Natural History: Can be observed on sandy or mud bottoms from low intertidal to depths of 200 feet. Common off Monterey. Phosphorescent; plankton feeder.

Size: Height to 61 cm.

Range: British Columbia to San Diego, California.

32. **ORANGE GORGONIAN** ***Adelogorgia phyllosclera***

(Shady-leafed, hidden gorgon)

Identification: The slender orange axis with yellow polyps has branches in a single plane.

Natural History: Can be observed attached to rocks in depths of about 100 feet to 1000 feet.

Size: Colony may reach 60 cm. in height.

Range: Southern California and Baja California.

33. **PURPLE GORGONIAN** ***Eugorgia rubens***

(Reddish true gorgon)

Identification: The slender purple or violet axis has white polyps; the branches form an interwoven pattern in one plane.

Natural History: Attaches to rocks in depths of 90 to 100 feet and deeper.

Size: May reach 180 cm. in height; however, most are 30 cm. or less.

Range: Southern California and Baja California.

30. FLESHY SEA PEN

31. WHITE SEA PEN

32. ORANGE GORGONIAN

33. PURPLE GORGONIAN

34. RED GORGONIAN — *Lophogorgia chilensis*

(Chilean crested gorgon)

Identification: The slender red axis with white polyps, branches in all directions.
Natural History: Colonies attached to rocks, particularly offshore pinnacles in depths of 50 to about 200 feet.
Size: To 90 cm. in height.
Range: Monterey Bay to the San Benitos Islands, Baja California.

35. BROWN GORGONIAN — *Muricea fructicosa*

(Fruitful purple one)

Identification: The colony, consisting of a thick brown axis, with white polyps, usually branches in one plane.
Natural History: One of the most common gorgonians in southern California; attaches to rocks or other solid substrate such as pier pilings, pipes, wrecks, etc., in depths of 50 to 100 feet. These gorgonians are capable of surviving in polluted inshore waters as well as in the clear, uncontaminated waters around offshore rocks.
Size: Colonies may reach 90 cm. in height.
Range: Point Conception to Cedros Island, Baja California.

36. CALIFORNIA GOLDEN GORGONIAN — *Muricea californica*

(California's purple one)

Identification: The colony consists of a thick brown axis with yellow polyps and branches in only one plane.
Size: Reaches 90 cm. in height.
Range: Point Conception to Baja California and into the Gulf of California.

37. SEA STRAWBERRY — *Gersemia rubiformis*

(Berry-shaped seed bearer)

Identification: The thick, bright red, soft lobes have white polyps.
Natural History: Colonies attach to offshore reefs and pinnacles from intertidal depths to about 50 feet. Sea strawberry colonies are one of the major foods of the large tochni nudibranch (*Tochuina tetraquetra*, #99).
Size: Reaches 10 cm. in height.
Range: Gulf of Alaska to Point Arena, California.

34. RED GORGONIAN

35. BROWN GORGONIAN

36. CALIFORNIA GOLDEN GORGONIAN

37. SEA STRAWBERRY

38. **PURPLE JELLYFISH** ***Pelagia panopyra***

(Fiery torch-of-the-sea)

Identification: The white medusa has purple radial markings.

Natural History: Free-swimming member of coastal plankton community. These large jellyfish serve as temporary homes for young crabs and at least one species of fish, the medusa fish (*Icichthys lockingtoni*). The purple jellyfish possesses very potent stinging cells (nematocysts) and should be avoided.

Size: Medusa (bell) may reach 45 cm. in diameter and, with trailing tentacles, may reach 450 to 600 cm. in length.

Range: Occurs throughout entire area at various times of the year.

39. **BROWN JELLYFISH** ***Chrysaora melanaster***

(Black star golden mouth)

Identification: The medusa is yellow-brown with 24 dark brown tentacles and a yellow-white manubrium (fleshy arms surrounding mouth).

Natural History: A very large member of the coastal planktonic community. The brown jellyfish possesses potent stinging cells and should be avoided. Like the purple jellyfish, the bell serves as a temporary home for young crabs and the medusa fish. Brown jellyfish feed on small fish, crustaceans and other members of the plankton community.

Size: Medusae may reach 30 cm. in diameter; the tentacles may reach to a length of 180 to 240 cm.

Range: Occurs at various times of the year in the entire area; usually winters along the California coast.

40. **COMMON JELLYFISH** ***Aurelia aurita***

(Golden pupa with ears [i.e. markings])

Identification: Lacks long tentacles and oral arms of *Pelagia* and *Chrysaora*. The translucent bell has horseshoe-shaped markings (gonads) around center and an eight-lobed margin.

Natural History: Common member of inshore plankton.

Size: Bell may reach 20 cm. in diameter.

Range: Occurs at various times along the entire Pacific coast.

PHYLUM ANNELIDA

POLYCHAETE WORMS

41. ***FEATHER*-DUSTER WORM** ***Eudistylia polymorpha***

(Clear column [with] many forms)

Identification: When this worm is exposed, the only visible feature is its generously frilled plume. Plume color varies from bright yellow to purple.

Natural History: *Eudistylia* lives only in rock areas from the intertidal to 80 feet. It builds its tube, which can be 25 cm. long, in narrow cracks and crevices. The plume serves two functions: respiration and food capture.

Size: Maximum plume diameter about 15 cm.

Range: Gulf of Alaska to San Diego, California.

38. PURPLE JELLYFISH

39. BROWN JELLYFISH

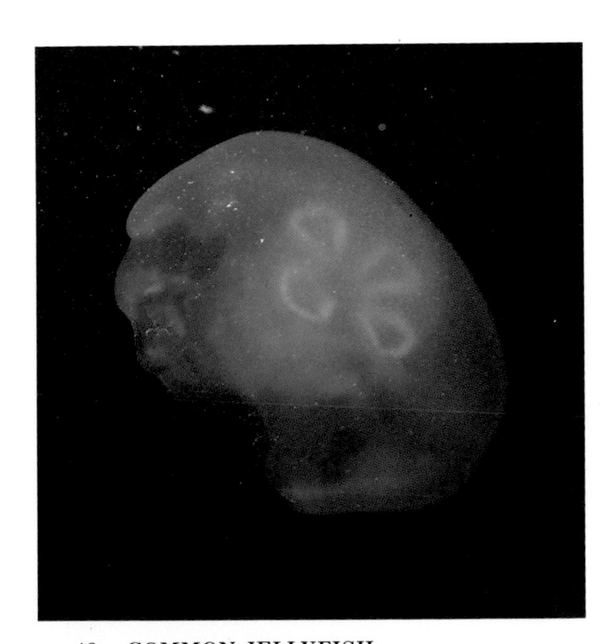

40. COMMON JELLYFISH

41. FEATHER-DUSTER WORM

42. **PLUME WORM** ***Serpula vermicularis***

(Wormy little snake)

Identification: The plumes are usually a bright red and, positioned between them, is a red-and-white, funnel-shaped operculum. Its shell is white, thickly calcareous and, usually tortuously coiled.

Natural History: *Serpula* grows exclusively on hard surfaces such as rock pilings and floats from the intertidal to 100 feet. Although the worm jerks back quickly into its tube when it perceives a threat, the operculum acts as a trap door and last line of defense.

Size: Maximum plume diameter to about 2 cm.

Range: Alaska to Baja California.

43. **FRAGILE TUBE WORM** ***Salmacina tribranchiata***

(Three-gilled fountain)

Identification: This tiny worm occurs in dense masses of intertwining tubes. The tubes are white and calcareous, but fragile. The worm is barely visible at the end of the tube as a smudge of red.

Natural History: *Salmacina* is much like *Eudistylia* (#41) or *Serpula* (#42) in function and design, but on a microscopic scale. It grows on rocks, generally in the protection offered by overhanging ledges. From the intertidal to 70 feet.

Size: Tube diameter about 0.25 cm.

Range: British Columbia to southern California.

44. **COLONIAL SAND TUBE WORM** ***Phragmatopoma californica***

(California hedge-cover)

Identification: This rarely visible worm posseses short, black tentacles and a black operculum which blocks the tube opening. The tube is composed of sand grains which the worm cements together.

Natural History: Never solitary, this worm can form large, honey-combed colonies. It occurs only on rock, but there must be an adequate amount of sand washed over the rock for the worms to capture, build and maintain their tubes.

From the intertidal to 240 feet.

Size: Individual tube diameter about 1 cm.

Range: Fort Bragg, California to Ensenada, Baja California.

45. **ORNATE TUBE WORM** ***Diopatra ornata***

(Adorned noble father)

Identification: The messy appearing tube is chitinized on the interior and covered with fragments of sticks, algae, shell and such which the worm applies to its exterior with specialized cement glands.

Natural History: Usually occurs in sand patches near rocks, either singly or in dense colonies. It is omnivorous, but utilizes algae to a large extent. From the shallow subtidal to 300 feet.

Size: Maximum diameter of tube about 1.5 cm.

Range: Central California to Western Mexico.

42. PLUME WORM

43. FRAGILE TUBE WORM

44. COLONIAL SAND TUBE WORM

45. ORNATE TUBE WORM

PHYLUM ARTHROPODA

BARNACLES, SHRIMP, CRABS, LOBSTER

46. **GIANT ACORN BARNACLE** ***Balanus nubilus***

(Cloudy acorn)

Identification: Size alone can distinguish the full grown adult from other white sessile or acorn barnacles; none approach it by half. Smaller specimens can be readily identified by the bright orange "lips" between the paired beaks. These fleshy "lips" are exposed when the beaks gape, especially during feeding.
Natural History: This barnacle requires a hard substrate, such as rocks or pier pilings, and strong currents or pounding waves. Intertidal to 100 feet.
Size: Maximum base diameter about 10 cm.; height to about 7.5 cm.
Range: Southern Alaska to San Quintin, Baja California.

47. **COON STRIPE SHRIMP** ***Pandalus danae***

(Dana's all-shining shrimp)

Identification: This very colorful shrimp has a red body interspersed with electric blue stripes. It lacks the large claws (chelae) of other shrimps, but has a very prominent upward curved rostrum. A closely related species, and difficult to distinguish, is *P. gurneyi* (not illustrated).
Natural History: The coon-striped shrimp can be found in bays and estuaries as well as the rocky subtidal. Intertidal to 600 feet.
Size: Maximum length to about 14 cm.
Range: Alaska to Monterey, California.

48. **NORTHERN KELP CRAB** ***Pugettia producta***

([crab] of Puget [Sound])

Identification: This kelp crab can be readily distinguished by its smooth shield-shaped carapace and long, needle-sharp walking legs. Its color is solid, red, brown or green, depending on habitat.
Natural History: This crab is generally associated with kelp. It often dwells in canopies formed by giant kelp, bull kelp or feather boa kelp, or among the lower growing red algae on the bottom. It is also commonly found on pier pilings and floating docks in protected bays. Intertidal to 250 feet.
Size: Greatest carapace width about 10 cm.
Range: British Columbia to Baja California.

49. **SOUTHERN KELP CRAB** ***Taliepus nutallii***

*(Nutalli's upon-high-one)

Identification: If its large size, stout long legs, and smooth, roundish carapace are not enough to identify this crab, its uniform bright red to dark purple color will certainly give it away.
Natural History: This crab can be found among kelp holdfasts and seaweeds on rocky bottoms in protected outer coasts from the shallow subtidal to 300 feet.
Size: Maximum carapace width to about 10 cm.
Range: Pt. Conception, California to Magdalena Bay, Baja California.
**Taliepus* is an anagram of *Epialtus*, its previous genetic name.

46. GIANT ACORN BARNACLE

47. COON STRIPE SHRIMP

48. NORTHERN KELP CRAB

49. SOUTHERN KELP CRAB

50. **MASKING CRAB** ***Loxorhynchus crispatus***

(Curly bent-nose)

Identification: This crab is distinguished by that which disguises it: its masking ability. One helpful feature used to identify this crab is its long, slim rostrum that diverges slightly at its bifurcated tips.

Natural History: Most often associated with rocks from the intertidal to 400 feet, masking crabs cover themselves with sponges, hydroids, bryozoans, tunicates and the like. From the intertidal to 400 feet.

Size: Maximum carapace width about 10 cm.

Range: Ft. Bragg, California to San Diego.

51. **SHEEP CRAB** ***Loxorhynchus grandis***

(Great bent-nose)

Identification: The sheep crab is probably the largest spider crab occurring within sport diving depths. It is recognized by its long legs and inflated, fuzzy-looking body. The rostrum is pitched steeply downward.

Natural History: *L. grandis* is a slow moving crab found on a wide variety of substrates, from 20 to 500 feet. It does not share the decorating genius of its smaller cousin, *L. crispatus* (#50).

Size: Maximum carapace width about 15 cm.

Range: Pt. Reyes, California to Baja California.

52. **ROCK CRAB** ***Cancer antennarius***

(Antenna crab)

Identification: This crab can usually be readily identified by its smooth, dark-reddish carapace contrasted by a yellowish, red-spotted underside. The walking legs are usually hairy. The black-tipped claws are really no help in identifying this animal as five *Cancer* species share this characterisitic: *C. gibbosulus*, *C. jordani*, *C. amphioetus*, *C. C. anthonyi* (#56) and *C. oregonensis*.

Natural History: This crab can be found in rocky habitat from the shallow subtidal to 100 feet or more. During daylight hours it usually is quiescent and hidden in cracks, crevices, holes, overhangs in rock, sand interfaces, etc.

Size: Maximum carapace width to about 13 cm.

Range: British Columbia to Magdalena Bay, Baja California.

53. **RED CRAB** ***Cancer productus***

(Extended crab)

Identification: The scientific name for this crab derives from the five teeth on the carapace between the eyes, which are more extended than in other cancer crabs to form a short rostrum. Color of adult body is red above to yellowish below. Juvenile coloration is often a pattern of bright zig-zag lines on the carapace.

Natural History: *Cancer productus* is another of the nocturnally active crabs and is found most often on rocky bottoms or on clean sand or cobble bottoms in bays. From the intertidal to about 300 feet.

Size: Maximum carapace width to about 17 cm.

Range: Alaska to Baja California.

50. MASKING CRAB

51. SHEEP CRAB

52. ROCK CRAB

53. RED CRAB

54. **DUNGENESS CRAB** ***Cancer magister***
(Master crab)

Identification: On the front edge of the carapace, between either eye and the widest point of the body, there are ten distinct, serrated projections (or "teeth") with the tenth "tooth" at the widest point of the crab. All other cancer crabs are widest at either the eighth or ninth tooth. Color is a light reddish brown.

Natural History: The Dungeness crab prefers a clean, sandy bottom from the shallow subtidal to about 1200 feet and is frequently obscured by total burial in the sand.

Size: Maximum carapace width to 20 cm.

Range: Unalaska to Baja California.

55. **SLENDER CRAB** ***Cancer gracilis***
(Slender crab)

Identification: The slender crab is often mistaken for a juvenile Dungeness crab (#54);

Natural History: As a larva and early juvenile, *Cancer gracilis* can be found on the underside of scyphozoan jellyfish, especially *Pelagia panopyra* (#38), the large white-and-blue jellyfish. Riding on the oceanic jellyfish, the crab achieves a distribution; the crab drops off the jellyfish when it reaches shallower nearshore waters. The adult crab is generally found on muddy or sand bottoms from the intertidal to 350 feet.

Size: Maximum carapace width to about 10 cm.

Range: Alaska to Baja California.

56. **YELLOW CRAB** ***Cancer anthonyi***
(Anthony's crab)

Identification: This crab resembles *Cancer antennarius* (#52), from which it can be distinguished by its more uniform yellowish color and by the lack of spotting on the underside. The yellow crab is also widest at its ninth carapace "tooth".

Natural History: This is the market crab of southern California, subject to a small, but growing, commercial fishery. Like the Dungeness crab, it prefers an offshore, sand bottom. In northern California its numbers appear to thin out and they are found most often in bays. The depth ranges from the shallow, subtidal to 300 feet.

Size: Maximum carapace width to 18 cm.

Range: Humboldt Bay to Playa Maria Bay, Baja California.

57. **HELMET CRAB** ***Telmessus cheiragonus***
(Little pond hand-clasper)

Identification: Living color varies from yellow-brown to dark red; the tips of the legs are often distinctly red while the legs themselves are hairy. The carapace has six large, serrated, triangular "teeth" on either side.

Natural History: Helmet crabs occur in both rocky and sandy habitats on the exposed coast and in bays. Found from the shallow subtidal to about 120 feet.

Size: Maximum carapace width about 10 cm.

Range: Siberia and North Pacific to northern California (Monterey).

54. DUNGENESS CRAB

55. SLENDER CRAB

56. YELLOW CRAB

57. HELMET CRAB

58. **MIMICKING CRAB** ***Mimulus foliatus***
(Leafy mimic)

Identification: The smooth, leaf-shaped carapace of this crab is about as wide as it is long. The mimicking crab often has sponges and bryozoans attached to the carapace. The color, often due to the encrusting sponges and/or bryozoans, can be either red or yellow.
Natural History: *Mimulus* occurs on rocky bottoms as well as sand from the low intertidal out to depths of about 75 feet.
Size: May reach 8 cm. in width.
Range: Alaska to Mazatlan, Mexico.

59. **SNOW CRAB** ***Chionoecetes bairdi***
(Baird's snow-dweller)

Identification: The claws, although well developed, are much shorter than the walking legs. Color of the living crab is yellowish and the spines and tubercules of the carapace are often red-tipped. The large recessed eye cups, visible from above, easily distinguish it from most other crabs.
Natural History: Divers in northern waters are most likely to encounter this crab in green or gray muds in water generally deeper than 30 feet. In the southern portion of its range, this crab inhabits deeper water. The maximum depth is about 1,500 feet.
Size: Greatest carapace width is about 14 cm.
Range: Bering Sea to Washington

60. **KING CRAB** ***Paralithodes camtschatica***
(Kamchatka's near stone)

Identification: King crabs have only three pairs of functional walking legs and both carapace and legs are covered with short spines. The color ranges from orange-yellow in juveniles to light purple in adults..
Natural History: King crabs prefer sand or mud bottoms in depths of about 10 to 600 feet. They migrate into shallow water during the spring to mate. King crabs, which may grow to be 30 years old, feed on clams, sea urchins and even fish.
Size: The maximum carapace width is about 30 cm.; however, the total width from leg tip to leg tip may reach 150 cm.
Range: North Pacific and Bering Sea south along the Pacific Coast to southeastern Alaska.

61. **PUGET SOUND KING CRAB** ***Lopholithodes foraminatus***
(Mandt's rock crest)

Identification: This scarlet or orange spinous crab is marked by prominent purple tubercles on its four pairs of legs and four tent-like projections on its carapace. When inactive, its legs fold neatly against its body, making it appear quite rock-like. A relative, *Lopholithodes foraminatus*, resembles it closely, but can be distinguished by a circular opening (foramen) formed between its claw legs and first walking appendage.
Natural History: Puget Sound king crabs do not occur in large aggregations. They are usually observed by divers on rocky reefs in depths as shallow as 20 feet out to depths of over 100 feet. The maximum depth is about 1800 feet.
Size: Carapace width to about 20 cm. (*L. foraminatus* = 30 cm. width)
Range: Sitka, Alaska to Monterey, California. (*L. foraminatus* = Alaska to San Diego, California).

58. MIMICKING CRAB

59. SNOW CRAB

60. KING CRAB

61. PUGET SOUND KING CRAB

62. **UMBRELLA CRAB** ***Cryptolithodes sitchensis***

(Sitka's cryptic, rock-like)

Identification: *Cryptholithodes* is difficult to confuse with any other crab as its carapace, when viewed from above, completely obscures the chelipeds and three pairs of walking legs.

Natural History: The umbrella crab is difficult to find without careful observation. It lies against rock or gravel backgrounds which it often exactly resembles. May be found from the intertidal to about 25 feet.

Size: Maximum carapace width about 7 cm.; length about 5 cm.

Range: Sitka, Alaska, to San Diego, California.

63. **HERMIT CRABS** ***Paguristes*** sp.

(A crab)

Identification: This crab is most easily recognized by its beggar's habit of borrowing the discarded homes of others—most often snail shells. Hermit crabs are divided into two families—those with equal sized chelae and those with unequal.

Natural History: Hermit crabs occur almost anywhere on the continental shelf from the intertidal to 1,200 feet. Their role in the food web is that of garbage collectors; they do the cleaning up of plant and animal debris.

Size: Maximum length of carapace (not including the soft, elongated abdomen) is about 2 cm.

Range: Bering Sea (Arctic Ocean) to Baja California

64. **SPINY LOBSTER** ***Panulirus interruptus***

(originally *Palinurus*)

(Sporadic backward tail)

Identification: The first thing a "bug" diver looks for are the long, sensitive antennae of this animal. Color varies from a dark black-red to a light red. Its body is armed with sharp spines for protection, especially above the eyes and along the tail.

Natural History: The spiny lobster prefers rocky areas with recesses and overhangs where it dwells during the day. Found from the intertidal to 200 feet.

Size: Maximum length over 60 cm.

Range: San Luis Obispo County, California, to Rosalia Bay, Baja California.

PHYLUM MOLLUSCA

CHITONS, LIMPETS, ABALONE, SNAILS
SEA HARES, NUDIBRANCHS, OCTOPUS, SQUID

65. **LINED CHITON** ***Tonicella lineata***

(lined extended chamber)

Identification: The lined chiton, to many observers, is the most beautiful chiton on this coast. With the brilliant, often blue, zig-zag pattern of lines on the background of red valves surrounded by a smooth, striped girdle it is readily distinguisable from the other chitons.

Natural History: The lined chiton occurs where crustose coralline algae forms hard red covering on rock. Intertidal to about 100 feet.

Size: Maximum length to about 50 cm.

Range: Aleutian Islands (Northern Japan) to San Diego.

62. UMBRELLA CRAB

63. HERMIT CRAB

64. SPINY LOBSTER

65. LINED CHITON

66. **PACIFIC GIANT CHITON** ***Cryptochiton stelleri***

(Steller's hidden chiton)

Identification: This animal more resembles a rugby ball, cut in half lengthwise, than it does a chiton. Its eight valves are completely covered by the fuzzy, brownish-red mantle. Visible from the underside are its mouth, large foot and gills lining the sides of the foot.

Natural History: Also known as the "Gumboot" chiton, this is the largest chiton in the world. Although it probably traverses sand in its migrations, its preferred habitat is rock where it grazes on algae. It may be found in the intertidal to 60 feet.

Size: Maximum length about 35 cm.

Range: Alaska to San Nicolas Island, California.

67. **GIANT KEYHOLE LIMPET** ***Megathura crenulata***

(Notched large window)

Identification: Its common and scientific names derive from the large oval hole at the apex of its shield-shaped shell. The shell is mostly internal, obscured by a soft fleshy mantle. The mantle is variously colored, ranging from jet black to a mottled yellow and green.

Natural History: *Megathura*, which is the largest of all the keyhole limpets on this coast, is generally found on exposed rocky reefs from the intertidal to about 110 feet. May feed on ascidian tunicates. Intertidal to 80 foot depths.

Size: Maximum length of 15 cm.

Range: Mendocino County to Baja California.

68. **ROUGH KEYHOLE LIMPET** ***Diodora aspera***

(Rough mountain passageway)

Identification: The conical shell of the rough keyhole limpet is completely exposed and its apical opening is nearly circular (vs. oval or elongate in other keyhole limpets). The shell sculpture is coarse and radial and generally a grayish-white color. Its foot is yellow.

Natural History: *Diodora* is found in rocky habitat on exposed outer coasts, usually among dense stands of algae, and ranges from the intertidal to 50 feet. The shell's apical hole serves as an exhaust port for exhaled water and wastes.

Size: Maximum length about 7 cm. (smaller in southern portion of range).

Range: Alaska to northern Baja California.

69. **WHITE CAP LIMPET** ***Acmaea mitra***

(Perfect cap)

Identification: Although there are at least 16 species of *Acmaeid* limpets, one of the most easily recognizable is *Acmaea mitra*. Also known as the "dunce-cap" limpet, its tall, conical shell, with its central apex, helps to distinguish it from its flatter relatives.

Natural History: The name "white-cap" limpet does not always fit older living specimens. The shell is often completely obscured by encrusting coralline or other red algae. This growth may serve as useful camouflage which protects the limpet from predators. It is found on algae-covered rocks from low intertidal to 100 feet.

Size: Maximum length of about 4 cm. and height of 3 cm.

Range: Aleutian Islands to northern Baja California.

66. PACIFIC GIANT CHITON

67. GIANT KEYHOLE LIMPET

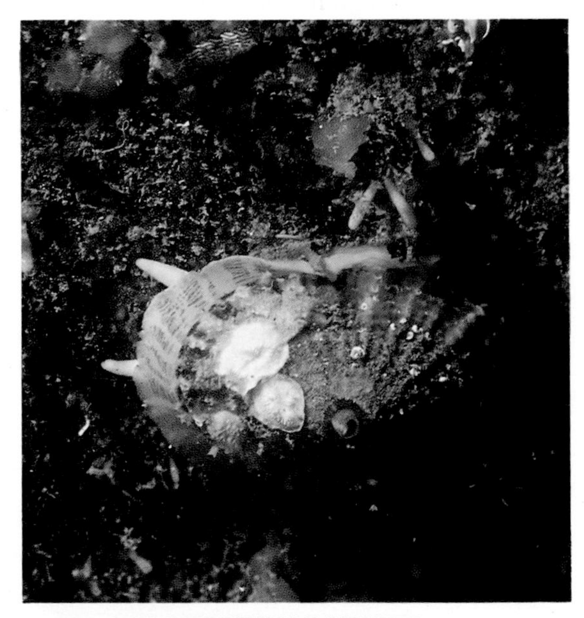

68. ROUGH KEYHOLE LIMPET

69. WHITE CAP LIMPET

70. **RED ABALONE** ***Haliotis rufescens***

(Red sea ear)

Identification: There are generally three to four open holes on the shell. While the shell exterior is most often a dull reddish-brown, it can also exhibit bands of pink, white or green. The shell interior, long prized for jewelry, is mother-of-pearl, with iridescent silver-greens and blues.
Natural History: Its shell is often concealed by a luxuriant growth of algae and animals, including barnacles, tunicates, hydroids, bryozoans and sponges. One sponge, *Cliona celata*, yellow in color, bores into abalone shells and often weakens them to the point of disintegration. Occurs from the intertidal to about 80 feet.
Size: Maximum length to about 28 cm. (minimum legal length 17.8 cm.)
Range: Coos Bay, Oregon to Baja California.

71. **BLACK ABALONE** ***Haliotis cracherodii***

(Cracherod's sea ear)

Identification: Its blue-green-black shell is the smoothest of all the abalone species on this coast. It also has five to eight open holes that are level with the top of the shell. The fleshy fringe on the foot, known as the epipodium, is also black.
Natural History: Although the black abalone lives primarily intertidally, it does occur in the very shallow subtidal. Its shell remains freer of fouling organisms than those of other abalone species. Intertidal to 20 feet.
Size: Maximum length about 17 cm. (minimum legal length 12.7 cm.)
Range: Mendocino, California to Baja California.

72. **FLAT ABALONE** ***Haliotis walallensis***

(Gualala sea ear)

Identification: The shell is quite flat; elongated and brick red in color—often with white, blue and green mottling. The shell has four to eight slightly raised open holes and the epipodium is lacy and yellowish-green with large brown-and-yellow marks.
Natural History: The flat abalone is found on and under rocks from about 20 to 70 feet and feeds on small algae. Its greatest predator, besides the sea otter, is the cabezon, *Scorpaenichthys marmoratus*.
Size: Maximum length to about 17.5 cm. (legal minimum is 10 cm.)
Range: British Columbia to Carmel Bay, rarely to La Jolla, California.

73. **PINTO ABALONE** ***Haliotis kamschatkana***

(Kamchatka's sea ear)

Identification: There are three to six (usually five) slightly raised open holes on this shell. Exterior shell color is a reddish-brown, mottled with white and blue markings.
Natural History: In the northern part of its range, this abalone occurs in the intertidal, but in the southern edge lives mostly in 35'-50' depths. Intertidal to 80 feet.
Size: Maximum length about 15 cm. (legal minimum is 10 cm.)
Range: Sitka, Alaska to Pt. Conception, California.

70. RED ABALONE

71. BLACK ABALONE

72. FLAT ABALONE

73. PINTO ABALONE

74. PINK ABALONE ***Haliotis corrugata***
(Wrinkled sea ear)

Identification: The shell, with two to four open holes projecting above its surface, has a sculpture of spiral and diagonal ridges which give it a corrugated appearance. Its exterior color is dull green to reddish-brown and the interior is strikingly iridescent in shades of green and pink. The lacy epipodium is mottled black and white.
Natural History: One of the most common abalone to be found around southern California's Channel Islands. Pink abalone prefer a shallow water, rocky habitat associated with beds of giant kelp. They have been taken at depths of 180 feet.
Size: Maximum size is about 25 cm. (legal minimum is 15.2 cm.)
Range: Pt. Conception to Turtle Bay, Baja California.

75. WHITE ABALONE ***Haliotis sorenseni***
(Sorensen's sea ear)

Identification: The shell has from three to five open holes on tubular projections. Shell color is generally a reddish brown and its surface sculpture is of diagonal undulations over a fine spiral ribbing. The interior is a pearly white with pale pink and green iridescence. The epipodium is lacy and mottled yellowish-green and beige.
Natural History: This southern abalone is rarely found shallower than 70 feet. Maximum depth is 200 feet.
Size: Maximum length is about 25 cm. (legal minimum size is 15.2 cm.)
Range: Pt. Conception to central Baja California.

76. GREEN ABALONE ***Haliotis fulgens***
(Shining sea ear)

Identification: The shell is relatively low and has five to seven moderately raised open holes. External color is brown to greenish-brown and the interior is an iridescent bluish-green. The epipodium is olive-green with brown patches and is scalloped along its edge.
Natural History: The green abalone dwells in shallow water on the open coast, the great majority being found in less than 30 feet. Low intertidal to 60 feet.
Size: Maximum length about 25 cm. (legal minimum is 15.2 cm.)
Range: Pt. Conception, California, to Baja California.

77. LEWIS' MOON SNAIL ***Polinices lewisii***
(Lewis' Polinices—son of Oedipus—the name means "many conquests")

Identification: This snail has a large grayish foot and a grayish mantle that, when extended, nearly covers its shell. The smooth brown shell is globular and has a large aperture with an umbilicus adjacent to it.
Natural History: This moon snail is found on sand-and-mud flats of protected bays and offshore in deeper water. It is a predator on bivalves (clams) and other snails. Shallow subtidal to 100 feet.
Size: Maximum diameter about 12 cm.
Range: British Columbia to northern Baja California.

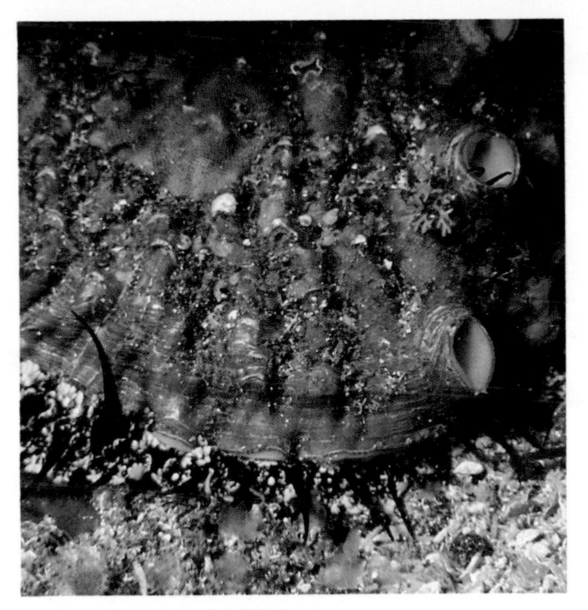

74. PINK ABALONE

75. WHITE ABALONE

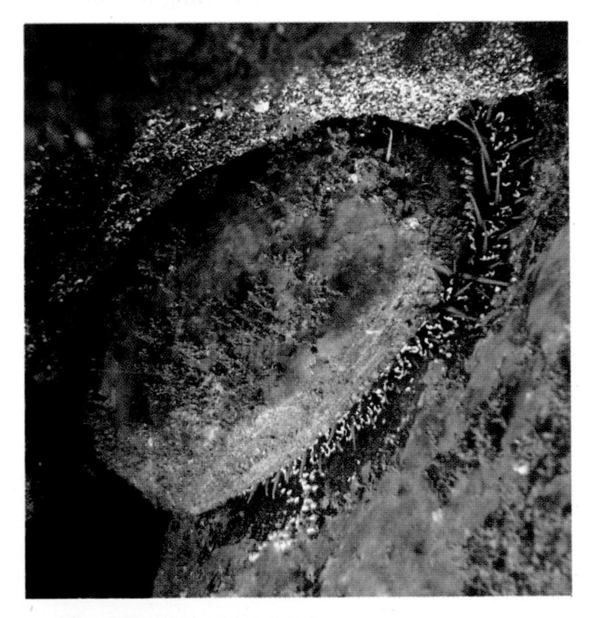

76. GREEN ABALONE

77. LEWIS' MOON SNAIL

78. **SMOOTH TURBAN** ***Norrisia norrisii***

(Norris's norris)

Identification: This snail has a very flattened spiral similar to the moon snail (#77). The dark red-brown, smooth shell has a pearly interior. The foot is a brilliant red in color.

Natural History: The smooth turban is an inhabitant of the kelp forest where it spends most of its life high above the sea floor. The depth range matches that of the giant kelp. Low intertidal to about 100 feet.

Size: Shell diameter to about 5 cm.

Range: Central California (rare) to Baja California.

79. **BROWN TURBAN SNAIL** ***Tegula brunnea***

(Brown roof or tile)

Identification: This snail has a black foot rimmed with a bright orange band. Its brown shell has slightly elevated whorls. When the snail is retracted within its shell, the membranous operculum, composed of concentric rings, is the only part that remains visible. The shell has one "tooth" on the inside of the aperture.

Natural History: *Tegula brunnea* is found in rocky habitats among dense stands of algae and in kelp forests among hold-fasts, and on kelp stipes. Mid-intertidal to about 40 feet.

Size: Maximum height to about 4 cm.

Range: Oregon to Southern California.

80. **RED TURBAN SNAIL** ***Astraea gibberosa***

(Humped star or "star maiden")

Identification: The background color of this conical shell is a light, reddish-brown. The base is marked by a strong spiral cord, and the heavy calcareous operculum is smooth and rounded.

Natural History: The red turban snail is found most often in rocky areas among dense coralline algae. Its shell is often completely covered by this calcareous algae. Intertidal to 100 feet.

Size: Maximum height about 5 cm.

Range: British Columbia to southern Baja California.

81. **WAVY TURBAN SNAIL** ***Astraea undosa***

(Wavy star)

Identification: This large snail's scientific and common names derive from the strong wavy sculpturing of the shell along the spiral ridge. Its shell is tan and is covered by a thick membrane known as a periostracum. It, too, has a thick, calcareous operculum that is marked by three strong, concentric ridges.

Natural History: The wavy turban occurs on rocky bottoms, especially in kelp beds. Shallow subtidal to 70 feet.

Size: Maximum height of 11 cm.

Range: Pt. Conception, California, to central Baja California.

78. SMOOTH TURBAN

79. BROWN TURBAN SNAIL

80. RED TURBAN SNAIL

81. WAVY TURBAN SNAIL

82. **BLUE OR COSTATE TOP SHELL** ***Calliostoma ligatum***

(Most beautiful banded mouth)

Identification: This snail is bluish-brown to reddish-brown, conical in outline, with elevated whorls and its sculpture is marked by clean, spiral ridges. When the outer surface erodes, a pearly layer is exposed. The body of the snail is black with a pinkish band around the foot.

Natural History: This top shell is found in rocky areas under ledges and among algae from the low intertidal to 60 feet. It is carnivorous, feeding mainly on sponges.

Size: Maximum height to about 3.5 cm.

Range: Alaska to San Diego.

83. **PURPLE-RINGED TOP SHELL** ***Calliostoma annulatum***

(Most beautiful ringed mouth)

Identification: The color alone is enough to identify this striking animal as no other snail comes close to its usually bright, gold-yellow background and its vivid purple band. Some specimens, however, appear faded and in these cases the straight pyramidal sides and beaded spiral ridges will aid in identification.

Natural History: This top shell is found in rocky habitats, often in kelp beds. Often found in association with, and presumably feeding on, the California hydrocoral, *Allopora californica*. Low intertidal to 100 feet.

Size: Maximum height is about 2.5 cm.

Range: Alaska to northern Baja California.

84. **CALIFORNIA CONE SNAIL** ***Conus californicus***

(California cone)

Identification: The color of this snail is a dull grayish-brown. Its most distinctive feature, besides the generalized cone shape, is the slit-like aperture occupying about 80% of shell length.

Natural History: It is found on rock and sand where it preys on worms and other snails. Intertidal to 150 feet.

Size: Maximum length is 4 cm.

Range: Farallon Islands, California, to southern Baja California.

85. **LEAFY HORNMOUTH** ***Ceratostoma foliatum***

(Leafy hornmouth)

Identification: A highly sculptured snail with three leafy, wing-like processes on the long axis of the shell and a tooth on the outside margin of the aperture. Shell color varies from white to light brown and it is sometimes banded.

Natural History: *Ceratostoma* is a predator and uses its tooth to feed on barnacles and nestling clams. It is generally found in exposed rocky areas and in kelp beds. Intertidal to 150-foot depth.

Size: Maximum length about 9 cm; average about 5 cm.

Range: Alaska to San Diego.

82, BLUE TOP SHELL

83. PURPLE-RINGED TOP SHELL

84. CALIFORNIA CONE SNAIL

85. LEAFY HORNMOUTH

86. **KELLET'S WHELK** ***Kelletia kelletii***

(Kellet's kellet)

Identification: Among spiral snails found in shallow water, **Kelletia** is one of the largest on this coast. Its heavy whitish shell has 8-9 rounded nodes per revolution.
Natural History: This is one of the more common snails in Southern California and is found on rocky and gravel bottoms. Depth from 10 feet to about 100 feet.
Size: Maximum height about 17 cm.
Range: Pt. Conception, California, to central Baja California.

87. **OREGON TRITON** ***Fusitriton oregonensis***

(Oregon Triton's spindle)
Triton= a sea god

Identification: This large snail has a brown shell with whorls decorated with 4-5 spiral lines of periostracal "hair". Each whorl is set off by a distinct shoulder and there is a tooth-like wrinkle at the top of the inner lip of the shell openng. Body color is red.
Natural History: One of the largest coiled snails to be found in our intertidal and shallow subtidal, out to depths of 7000 feet. It is found on both rock and sand substrate and feeds on a wide variety of worms, sea urchins and molluscs.
Size: Maximum length about 13.5 cm.
Range: Alaska to La Jolla.

88. **CHESTNUT COWRY** ***Cypraea spadicea***

(Nut-brown "Kyprio"—named for Venus or Aphrodite, hence: "love shell")

Identification: Unmistakable for any other snail, the highly polished shell is whitish at the base with an irregular chestnut-brown patch at its center. If undisturbed, the shell is often covered by a highly papillated mantle which keeps the shell shiny.
Natural History: It generally occurs on rocky substrate, quite often on offshore pinnacles where there is good water circulation. It feeds on sponges, tunicates, even small sea anemones. Low intertidal to 100 feet.
Size: Maximum length to 6.5 cm.
Range: Monterey, California, to central Baja California.

89. **IDA'S MITER** ***Mitra idae***

(Ida's miter)

Identification: The elongate shell is smoothly tapered to its apex and is dark-brown to black in color. There are three to four folds in the white inner part of the aperture. Body color is white and its tubular siphon is often exposed.
Natural History: *Mitra* occurs in rocky areas, especially in kelp beds from the low intertidal to 70 feet. Once divided into several species, based largely on length, *Mitra idae* now combines both long and short forms under its nomenclatural roof.
Size: Maximum length to 6 cm.
Range: Crescent City, California, to central Baja California.

86. KELLET'S WHELK

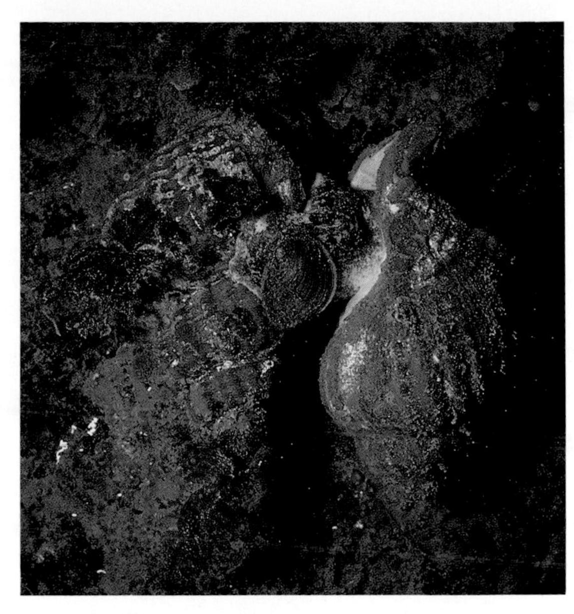

87. OREGON TRITON

88. CHESTNUT COWRY

89. IDA'S MITER

90. **SCALED WORM SHELL** ***Serpulorbis squamigerus***

(Circular, scale-bearing, little snake)

Identification: The whitish-gray tube is longitudinally ribbed, and the snail has a black body ringed with orange at the circular aperture of the tubes. The tube shape is various—sometimes straight but most often twisted.

Natural History: North of Pt. Conception, *Serpulorbis* grows singly, firmly attached to rock substrate but in the southern part of its range the tubes grow into connected, tortuously twisted masses often forming large colonies. Low intertidal to 100 feet.

Size: Tube diameter about 1 cm.

Range: Pt. Sur area (California) to central Baja California.

91. **SEA HARE** ***Aplysia californica***

(California filthy [one])

Identification: This large, shell-less mollusc is mottled tan, olive-green, red or olive-brown in color. Its common name comes from the two appendages (rhinophores) behind its head on its back which resemble rabbit's ears. On its back are two fleshy flaps which are extensions of the foot.

Natural History: *Aplysia* is found in both rocky and sandy offshore habitat, as well as on protected mudflats, from the intertidal to 80 feet.

Size: Maximum length about 40 cm., weight to 30 pounds.

Range: Trinidad Bay, California, to northern Baja California.

92. **ORANGE-TIPPED NUDIBRANCH** ***Triopha catalinae***

(Catalina's sea nymph)

Identification: The ground color of this dorid nudibranch is grayish-white, although most of the dorsal processes and tubercles are tipped with bright orange. The species is marked by a frontal veil composed of 8-12 orange lobes.

Natural History: *Triopha* is commonly found in rocky habitats where wave energy is moderately strong. It is also found in bays on docks and pilings. It feeds exclusively on branching bryozoans. Found from the intertidal to depths of about 100 feet.

Size: Maximum length about 5 cm.

Range: Alaska to San Diego, California.

93. **WHITE-SPECKLED NUDIBRANCH** ***Dendrodoris albopunctata***

(White speckled, tree-like Doris [a sea goddess])

Identification: The ground color of this dorid varies from a pale whitish yellow to a dark yellow-orange. On darker specimens the dorsal white speckling stands out vividly. Its gill tree is white to pale yellow.

Natural History: This is one of the most common dorid nudibranchs in Central California. It is found on rocks in exposed areas and in kelp beds and it feeds on sponges. Found out to depths of about 140 feet.

Size: Maximum length to about 4 cm.

Range: Mendocino County, California, to Baja and northern Gulf of California.

90. SCALED WORM SHELL

91. SEA HARE

92. ORANGE-TIPPED NUDIBRANCH

93. WHITE-SPECKLED NUDIBRANCH

94. **YELLOW-EDGED CADLINA** ***Cadlina luteomarginata***

(Yellow-edged little cask)

Identification: The general ground color of this dorid is a semi-transluscent white and on the back are numerous tubercles tipped in bright yellow. A similarly bright yellow band decorates the entire perimeter of the mantle.
Natural History: This easily recognized dorid is found in rocky habitats from the intertidal to 110 feet. It feeds on a variety of sponges.
Size: Maximum length about 4.5 cm.
Range: Auke Bay, Alaska, to Point San Eugenio, Baja California.

95. **RINGED NUDIBRANCH** ***Discodoris sandiegensis***

(San Diego's disk-shaped Doris [a sea goddess])

Identification: The most common background color of this animal is a creamy white, although it can vary from off-white to light brown. The back is generally marked with 2-8 brown-to-black rings which are often paired but can also be irregular in design and more numerous.
Natural History: This wide-ranging dorid is found commonly in both rocky habitats on the exposed open coast to depths of 100 feet, and on rocks, pilings and docks within bays. It feeds on a wide variety of sponges.
Size: Maximum length to about 15 cm., average closer to 3.5 cm.
Range: Unalaska, Alaska, to Cabo San Lucas,Baja California.

96. **FALSE SEA LEMON** ***Archidoris montereyensis***

(First Doris [of] Monterey)

Identification: *Archidoris* strongly resembles *Anisodoris nobilis* (#98). Its ground color ranges from light yellow to orange-brown and its back is spattered with brown brown to black dots which extend onto the prominent tubercles. Its branchial plumes are yellow to orange in color.
Natural History: This is one of the more common dorid nudibranchs and it is found in rocky habitat on outer coasts to depths of 800 feet, as well as in bays on docks and mud. It feeds on sponges.
Size: Maximum length to 15 cm., average closer to 3.5 cm.
Range: Port Valdez, Alaska, to San Diego, California.

97. **WHITE SEA LEMON** ***Archidoris odhneri***

(Odhner's first Doris [a sea goddess])

Identification: Apart from its great size, this dorid is readily identified by its snow white color and by the large tubercles covering its back.
Natural History: The largest dorid nudibranch on this coast. It occurs fairly commonly in rocky areas in and around kelp beds from the intertidal to depths of 80 feet. It feeds on a wide assortment of sponges.
Size: Maximum length about 20 cm., average about 10 cm.
Range: Auke Bay, Alaska, to San Diego, California.

94. YELLOW-EDGED CADLINA

95. RINGED NUDIBRANCH

96. FALSE SEA LEMON

97. WHITE SEA LEMON

98. **SEA LEMON** ***Anisodoris nobilis***

(Well-known, unequal Doris [a sea goddess])

Identification: This common bright yellow nudibranch has prominant, knob-like tubercles on its back with brown to black spots interspersed *between* the tubercles. The branchial plume is light yellow.

Natural History: One of the largest and most common nudibranchs on our coast. The sea lemon has a fruity, penetrating odor similar to other dorid nudibranchs. Sea Lemons are hermaphoditic; the strings of light yellow eggs are deposited during the winter. They occur from the intertidal out to 300 feet.

Size: The average size is about 10 cm., but 20 cm. sea lemons have been documented.

Range: Vancouver Island to Baja California.

99. **TOCHNI NUDIBRANCH** ***Tochuina tetraquetra***

(Tochni's [Indian name] four-sided-one)

Identification: The general ground color of this nudibranch is orange-yellow to brownish-yellow and its back is covered with many white-tipped tubercles. The margin is lined with a series of white plumose branchial appendages.

Natural History: This large and striking eolid is found on both rocky areas and sandy-mud bottoms in depths from 20 to 1000 feet. It feeds on both strawberry coral (*Gersemia rubiformis*) on rock habitat and on sea pens (*Ptilosarcus gurneyi*) on flat sand bottoms.

Size: Maximum length to 30 cm.; average about 12 cm.

Range: Alaska to Santa Barbara Island.

100. **TRITONS NUDIBRANCH** ***Tritonia festiva***

(Gay colors [of] Triton)

Identification: This elongate nudibranch is characterized by a series of reticulate, rather diamond-shaped, white lines on the back. Ground color is white to yellowish-white and a series of 11-15 tree-like branchial plumes is located along both sides of the upper body margin. It also possesses a well defined frontal veil.

Natural History: This small nudibranch occurs on both rock and sand from the intertidal to 60 feet. It feeds on an octocoral (*Clavularia* sp.), a sea pen (*Ptilosarcus*) and a gorgonian coral (*Lophogorgia*).

Size: Maximum length about 3 cm.

Range: Vancouver Island, British Columbia, to northern Baja California.

101. **RAINBOW NUDIBRANCH** ***Dendronotus irus***

(Rainbowed tree-back)

Identification: This nudibranch has a series of 4-7 pair of arborescent processes along its dorsal margin. The rhinophore sheath is also tree-like. There are two general color phases: translucent gray-white background color with white tipped cerata and a dark red ground color with orange tipped cerata.

Natural History: This nudibranch is found on flat sand bottoms from the intertidal to 650 feet, but most commonly at 30-40-foot depths where it feeds on the tube anemone (#26).

Size: Maximum length to 20 cm.; average about 6 cm.

Range: Unalaska, Alaska, to northern Baja California.

98. SEA LEMON

99. TOCHNI NUDIBRANCH

100. TRITONS NUDIBRANCH

101. RAINBOW NUDIBRANCH

102. **WHITE-LINED NUDIBRANCH** ***Dirona albolineata***

(White-lined, dreadful burden)

Identification: The ground color of this nudibranch is a translucent white and on its back it bears many pointed, flattened cerata with an opaque white line on their margin.
Natural History: *Dirona* occurs in rocky areas in the open, exposed coast. It is apparently an undiscerning predator with an unusually wide-based appetite, reportedly feeding on hydroids, bryozoans, sponges, barnacles, snails and tunicates. Intertidal to 100 feet.
Size: Maximum length to 18 cm.; average about 3 cm.
Range: Vancouver Island, British Columbia, to San Diego, California.

103. **HILTON'S NUDIBRANCH** ***Phidiana hiltoni***

(Hilton's Phidias [a Greek Sculptor])

Identification: The long, tapering body of this aeolid is densely covered with dark cerata having white tips. There are about 30 rows of cerata along either side of the back and they are divided into two or three distinct regions. Body color is white and there is generally an orange stripe between its tentacles on the head, as well as orange bonding on the rhinophores.
Natural History: *Phidiana* occurs on rock substrate from the intertidal to depths of 650 feet. It feeds on hydroids.
Size: Maximum length to about 5 cm.
Range: Monterey Bay, California, to Guardian Angel Island, Gulf of California.

104. **HORNED NUDIBRANCH** ***Hermissenda crassicornis***

(Thick-horned)

Identification: The general ground color of this animal is translucent grayish-white. It is decorated with opalescent blue lines—one borders either side of the foot and another runs down the midline. Also on the midline are two bright gold linear marks—one between the rhinophores and the other at mid-body. The cerata are long and numerous, brownish-yellow in color, and white-and-gold tipped.
Natural History: It is found in almost every habitat in depths from the intertidal to 120 feet. It is especially abundant on rocks in protected bays. It feeds on a wide variety of hydroids.
Size: Maximum length about 5 cm.
Range: Sitka, Alaska, to Guardian Angel Island, Baja California.

105. **SPANISH SHAWL** ***Flabellinopsis iodinea***

(Appearing like a violet fan)

Identification: The body color of this beautiful aeolid is bright purple and the back bears many cerata on longitudinal ridges. The cerata, purple at their base, are a brilliant orange for most of their length.
Natural History: *Flabellinopsis* occurs on rocky habitat along the open coast from the intertidal to 120 feet and occasionally on pilings and docks in bays.
Size: Maximum length about 5 cm.
Range: Vancouver Island, British Columbia, to Cape San Quintin, Baja California.

102. WHITE-LINED NUDIBRANCH

103. HILTON'S NUDIBRANCH

104. HORNED NUDIBRANCH

105. SPANISH SHAWL

106. **ROCK SCALLOP** ***Hinnites giganteus***
(Giant mule)

Identification: The color of the exterior of an unfouled (i.e., by encrusting organisms) shell is yellowish-orange; the interior is white and on the hinge of both valves is a purple stain. Attached to rock by one valve as an adult, the shell is usually highly irregular.
Natural History: It is found in crevices and on walls in rocky habitats from the intertidal to 100 feet, especially on offshore reefs and pinnacles where water movement is greater; also in bays or pier pilings and on rocks.
Size: Maximum diameter about 25 cm.; average about 12.5 cm.
Range: British Columbia to central Baja California.

107. **ABALONE JINGLE** (or **ROCK OYSTER**) ***Pododesmus cepio***
(Ligament-footed onion)

Identification: This nearly round shell is composed of two unequal valves. The exposed uppermost valve has irregular radial ribs and is grayish-white in color. The lower valve is more wrinkled and has a notched hole near the hinge through which the animal's attachment organ holds it to the substrate. The interior color of the shell is pearlescent green-gray.

Natural History: They occur on rocks and pilings in bays and estuaries from the intertidal to 100 feet. The shell is difficult to see because of its low profile and because it is often overgrown by worms, barnacles, sponges, etc.
Size: Maximum diameter to about 10 cm.
Range: British Columbia to Gulf of California.

108. **GAPER CLAM** ***Tresus nuttallii***
(Nuttall's hole)

Identification: Only the siphon is seen by divers. These are recognizable by the large openings (about 1.5 cm. on large specimens) of the fused incurrent and excurrent siphons. The tip of the siphon's exterior is a pair of thin protective siphonal plates which are often overgrown by small, attached invertebrates and algae.
Natural History: The gaper clam is found in sandy muds of estuaries and protected bays. It resides in a fairly deep burrow (2-3 feet deep).
Size: Maximum shell length about 25 cm.
Range: Washington state to Baja California.

109. **HEART COCKLE** (or **BASKET COCKLE**) ***Clinocardium nuttallii***
(Nuttall's sloping heart)

Identification: This cockle belongs to a family of clams (Cardiidae) that gets its name from the heart-shaped profile of its two shells (when viewed from the end). The shell color of the heart cockle is usually brownish-gray to white and each valve has 35 to 37 strong, radiating ribs. The siphon is very short, barely extending beyond the shell.
Size: Maximum length about 10 cm.
Range: Bering Sea to northern Baja California.

Natural History: It occurs predominantly in mudflats of bays, but can also be found in clean sand in offshore areas from the mid-intertidal to 50 feet.

106. ROCK SCALLOP

107. ABALONE JINGLE

108. GAPER CLAM

109. HEART COCKLE

110. **BORING CLAM** ***Pholadidae***

(A family name: Lurking in a hole)

Identification: Living *Pholads* are only evident by the tips of their siphons which are either even with or protrude slightly from the substrate. The animal pictured here is **the wart-necked piddock, *Chaceia ovoidea*.**

Natural History: *Pholad* clams, also called "piddocks", are found in sedimentary rock and hard muds from the intertidal to 150 feet. The hole is created by a mechanical rasping of the file-like shell rotating against the substrate.

Size: Shell length to 12 cm. The diameter of the individual siphon may reach 2 cm.

Range: Alaska to Baja California.

111. **OCTOPUS** ***Octopus*** spp.

(eight legs)

Identification: All octopuses are mimickers, readily changing colors and, often, even skin texture.

Natural History: The octopus is generally considered to be the most intelligent of all invertebrates. By day it usually hides under or among rocks and can also be found in discarded cans and bottles. It feeds primarily on crustaceans. Intertidal to several hundred fathoms.

Size: Maximum arm spread about 300 cm.; average closer to 30 cm.

Range: Entire Pacific coast.

112. **MARKET SQUID** ***Loligo opalescens***

(Opalescent squid)

Identification: The squid is an elongate animal with two wing-like extensions near the apex and ten arms. It shares with the octopus the ability to instantly change its body coloring and patterning with subcutaneous organs called chromatophores.

Natural History: About the only time a diver or sport fisherman will see a squid is when it migrates from deep, offshore water to mate and to spawn in shallow near-shore waters, usually in semi-protected bays. Depths range from 30 feet to mid-oceanic depths.

Size: Maximum length to about 40 cm.

Range: British Columbia to Baja California—perhaps to the Gulf of California.

PHYLUM ECTOPROCTA

BRYOZOANS

113. **FLUTED BRYOZOAN** ***Hippodiplosia insculpta***

(Carved + double-horse)

Identification: Look for the yellow to orange upright "curled-leaf-shaped" colonies.

Natural History: Occurs attached to rocks or kelp from the low intertidal out to depths of 700 feet. Most common on reefs where currents are strong enough to provide large amounts of planktonic food.

Size: Colonies reach 10 cm. in height and 10 to 12 cm. in diameter.

Range: Gulf of Alaska to Sacramento Reef, Baja California.

110. WART-NECKED PIDDOCK

111. OCTOPUS

112. MARKET SQUID

113. FLUTED BRYOZOAN

114. **LACY BRYOZOAN** ***Phidolopora pacifica***

(Pacific sparse pores)

Identification: The colonies are orange to orange-brown, delicate, interwoven and lace-like.
Natural History: Lacy bryozoan colonies can be observed attached to rocks, particularly those offshore pinnacles and reefs where large amounts of planktonic food is available. Depths range from low intertidal to 600 feet.
Size: Colonies reach 12 cm. in height and 18 or 20 cm. in diameter.
Range: Gulf of Alaska to Cedros Island, Baja California.

115. **NORTHERN STAGHORN BRYOZOAN** ***Heteropora magna***

(Great different pore)

Identification: The buff to brown, club-like, upright branches of the colony have an almost round cross-section.
Natural History: The coral-like colonies attach to offshore rocky reefs where plankton food is more abundant. We are aware of this bryozoan occurring only in depths of about 40 to 90 feet.
Size: Colonies reach about 10 cm. in height and about 15 cm. in diameter.
Range: Gulf of Alaska to Point Buchon, California.

116. **SOUTHERN STAGHORN BRYOZOAN** ***Diaperoecia californica***

(California perforated dwelling)

Identification: The light-yellow branches have a flat cross-section.
Natural History: Colonies occur on offshore reefs, as well as intertidal rocks. The coral-like colonies have been collected from as deep as 600 feet.
Size: Reaches a height of 10 cm. and a diameter of 13 or 15 cm.
Range: Salt Point, California, to San Benitos Islands, Baja California.

PHYLUM ECHINODERMATA

SEA STARS, URCHINS, CUCUMBERS, BRITTLE STARS

117. **SLATE PENCIL URCHIN** ***Eucidaris thauarsii***

(Thauar's true turban)

Identification: This distinctive urchin has thick, blunt-tipped, pencil-like spines.
Natural History: Usually observed in crevices in rocky reefs from the intertidal out to depths of 500 feet. Rare in California.
Size: May reach 20 cm.
Range: Santa Catalina Island to Ecuador and Galapagos Islands.

114. LACY BRYOZOAN

115. NORTHERN STAGHORN BRYOZOAN

116. SOUTHERN STAGHORN BRYOZOAN

117. SLATE PENCIL URCHIN

118. **GIANT RED SEA URCHIN** ***Strongylocentrotus franciscanus***

(Franciscan round [ball] of spines)

Identification: The length of the long, slender, smooth spines is about equal to the test in diameter. Color varies from red to dark purple.
Natural History: Common to abundant on rocky reefs from the low intertidal out to depths of at least 100 feet. Giant red sea urchins feed primarily on kelp and other algae, but can also subsist on animal matter. They, in turn, are a favored food of sea otters. Gonads are considered a delicacy by some.
Size: Test reaches 20 cm. in diameter.
Range: Gulf of Alaska to Cedros Island, Baja California.

119. **CORONADO SEA URCHIN** ***Centrostephanus coronatus***

(Wreathed spiny-crowned [urchin])

Identification: This urchin has very long (at least twice test diameter), serrated (to touch) spines. Color is uniform dark purple.
Natural History: This long-spined urchin is usually found in rock crevices during the day, but can be observed moving about in the open at night in search of food. The depth range is from the low intertidal out to depths of 350 feet.
Size: Test reaches 15 cm. in diameter, with spines about twice as long as test diameter.
Range: Southern California to Baja California and the Gulf of California.

120. **PURPLE SEA URCHIN** ***Strongylocentrotus purpuratus***

(Purple [ball] of-spines)

Identification: Look for the short, bright blue to purple spines in adults; in juveniles (less than one inch in diameter) the spines are greenish.
Natural History: Occurs on rocky bottoms from the intertidal out to a depth of 80 feet. Purple urchins are abundant in areas where there is heavy surf. They use their spines to burrow into rocks and when an urchin dies, its burrow becomes a home for another animal, such as a crab.
Size: Reaches 8 cm. in diameter.
Range: Alaska to Cedros Island, Baja California.

121. **GREEN SEA URCHIN** ***Strongylocentrotus droebachiensis***

(Droebach's ball of spines)

Identification: The green spines are usually shorter in length than the diameter of the test.
Natural History: Occurs on rocky bottoms from the low intertidal out to depths of 50 or more feet. Feeds on animal matter as well as algae.
Size: Reaches about 10 cm. in diameter.
Range: Gulf of Alaska to Puget Sound Washington. Also found in North Atlantic.

118. GIANT RED SEA URCHIN

119. CORONADO SEA URCHIN

120. PURPLE SEA URCHIN

121. GREEN SEA URCHIN

122. **WHITE SEA URCHIN** ***Lytechinus pictus***

Painted broken urchin)

Identification: The white spines are usually equal to or shorter than the diameter of the test.
Natural History: Occurs on sand, mud and rock bottoms in depths of 20 to 100 feet. From our observations, they appear to be scavengers.
Size: Maximum diameter about 5 cm.
Range: Santa Barbara, California to Thurloe Head, Baja California.

123. **SAND DOLLAR** ***Dendraster excentricus***

(Off-center tree-star)

Identification: Living animals are dark purple with very short and dense light-colored spines. The typical "five-leaf" pattern seen on dead tests is not readily apparent in living animals.
Natural History: Sand dollars are really flat urchins which occur partially buried in shallow sand bottoms in depths of 15 to 50 feet. Sand dollars feed on diatoms attached to sand grains.
Size: Reaches 10 cm. in diameter.
Range: Alaska to Baja California.

124. **SUNFLOWER STAR** ***Pycnopodia helianthoides***

(Many-legged sunflower)

Identification: This star has 15 to 24 limp, flexible arms with a fleshy covering. The color varies from purple to various shades of brown.
Natural History: This common and distinctive sea star can be observed on rocks, sand and mud bottoms from the intertidal out to depths of over 100 feet. A voracious predator, it feeds on abalone, other sea stars, urchins and many other animals.
Size: May reach 90 cm. in diameter.
Range: Alaska to San Diego, California.

125. **ORANGE SUN STAR** ***Solaster stimpsoni***

(Stimpson's sun-star)

Identification: The 10 to 12 rays have a blue-gray stripe, bordered by pink, red or orange, extending from the disc out to the tip of each arm.
Natural History: Occurs from the intertidal out to 1080 feet on rocky bottoms.
Size: May reach 50 cm. in diameter.
Range: Alaska to Salt Point in Sonoma County, California.

122. WHITE SEA URCHIN

123. SAND DOLLAR

124. SUNFLOWER STAR

125. ORANGE SUN STAR

126. **MORNING SUN STAR** ***Solaster dawsoni***

(Dawson's sun-star)

Identification: This eight to 15-armed star has a uniform color of orange, dull yellow, brown or blue-gray. Flat-topped spinelets on ventral surface of arms are larger than those of the orange sun star.
Natural History: The morning sun star has been observed on rocky bottoms from the shallow subtidal out to 200 feet. The morning sun star occasionally feeds on the orange sun star and other sea stars.
Size: Maximum size is between 38 and 50 cm. in diameter.
Range: Alaska to San Diego, California.

127. **ROSE STAR** ***Crossaster papposus***

(Downy fringed star)

Identification: This star has a large disc with clusters of spines resembling tiny brushes on an open network of calcareous ridges and eight to 14 short arms whose length beyond edge of the disc does not exceed the diameter of the disc.
Natural History: Rarely found intertidally. Common on subtidal rocks out to depths of 2200 feet.
Size: May reach 25 cm. in diameter.
Range: A circumpolar sea star known to occur on our coast as far south as Puget Sound.

128. **SIX-RAYED STAR** ***Leptasterias hexactis***

(Six-rayed + small star)

Identification: The six arms are covered with short spines; mottled red and gray in color.
Natural History: The six-rayed star can be found on rocky bottoms from the intertidal out to depths of 684 feet. Females carry eggs until young sea stars are released.
Size: Maximum size is from 7.5 to 10 cm. in diameter.
Range: Alaska to Point Conception, California.

129. **BLOOD STAR** ***Henricia leviuscula***

(Henry's smooth-little-one)

Identification: Look for five stiff, slender and smooth arms lacking spines and pedicellaria; color in adults is usually bright red to orange.
Natural History: Occurs on rocks from the intertidal zone out to 2200 feet. This colorful sea star also broods the eggs—the female carries the eggs until the young sea stars are born.
Size: Maximum size is from 20 to 25 cm. in diameter. The largest individuals we have observed were at Kodiak Island.
Range: Aleutian Islands to central Baja California.

126. MORNING SUN STAR

127. ROSE STAR

128. SIX-RAYED STAR

129. BLOOD STAR

130. **FRAGILE STAR** ***Linckia columbiae***

(Linck's dove)

Identification: A smooth gray and red mottled star with legs usually of unequal length and nearly round in cross-section. May have more than five arms.
Natural History: Common on rocky reefs in southern California, particularly offshore islands; from intertidal depths out to 240 feet. Rarely found in symmetrical five-arm condition and little is known regarding the reasons for the ease with which this mottled sea star breaks off arms (autotomy). They may lose arms during changing environmental conditions. Experiments indicate it requires about six months for a new disc and arms to grow from a separated arm. Another interesting fact is that a single animal may possess as many as five mouths.
Size: Reaches at least 10 cm. in diameter.
Range: San Pedro, California to Galapagos Islands.

131. **SEA BAT** ***Patiria miniata***

(Vermillion dish)

Identification: A five-armed (occasionally four to nine) sea star with indistinct separation of arms with a scaled surface. The color is highly variable, from solid red to mottled yellow, orange or brown.
Natural History: Common in central and southern California on sand and mud bottoms, as well as on rocky reefs. Their depth range extends from the intertidal out to 90 feet. Eggs and sperm are discharged throughout the year. Sea bats tend to be scavengers and will congregate in large numbers in areas where other animals are dying.
Size: Reaches about 15 cm. in diameter.
Range: Sitka, Alaska, to San Benitos Islands, Baja California.

132. **LEATHER STAR** ***Dermasterias imbricata***

(Tiled + leather star)

Identification: The only five-armed sea star in our area that has a soft, smooth to slippery upper (dorsal) surface. The color is mottled gray-green and reddish brown.
Natural History: Leather stars can be observed on rocky reefs as well as on sand bottoms in depths to 100 feet.
Size: Reaches at least 25 cm. in diameter.
Range: Gulf of Alaska to Sacramento Reef, Baja California.

133. **RED SEA STAR** ***Mediaster aequalis***

(Equal + middle star)

Identification: A bright red, five-armed star with small plates along margin of arms.
Natural History: Occurs on rocky reefs in depths of 50 to 1650 feet. Common in Carmel Bay, California on encrusting coralline algae.
Size: Diameter reaches at least 10 cm.
Range: Alaska peninsula to Baja California.

130. FRAGILE STAR

131. SEA BAT

132. LEATHER STAR

133. RED SEA STAR

134. **FAT SEA STAR** ***Pteraster tesselatus arcuatus***

(Curved, checkered wing-star)

Identification: The broad, thick, short arms are the best characteristic to use for identification. Color varies from mottled gray and black to brown and orange.
Natural History: This sea star occurs on rocky reefs in depths of 18 feet (Alaska) to 1100 feet. They are usually not found any shallower than 50 feet off California.
Size: Reaches 15 cm. in diameter.
Range: Bering Sea to Carmel Bay, California.

135. **SPINY SEA STAR** ***Poraniopsis inflata***

(Swollen pore-anus)

Identification: A five-armed star with numerous long spines in margins of arms and on dorsal surface. The color is usually orange.
Natural History: This deep-water near relative of the blood sea star occurs on rocky reefs in depths of 60 to 960 feet. Spiny sea stars probably prey on snails and other molluscs.
Size: May reach 15 cm. in diameter.
Range: British Columbia to San Diego, California.

136. **SPINY SAND STAR** ***Astropecten armatus***

(Armed star-comb)

Identification: A very symmetrical five-armed sea star with conspicuous plates on margins of arms. The color is usually gray or tan.
Natural History: Occurs on sand and mud bottoms in depths ranging from low intertidal to 30 fathoms.
Size: May reach 25 cm. in diameter.
Range: Channel Islands, California, south to Ecuador. A similar species, *Astropecten verrilli* (not illustrated), ranges as far north as Point Reyes and is common in Monterey Bay.

137. **SAND STAR** ***Luidia foliata***

(Leafy flat star)

Identification: A star with five long, narrow and smooth arms. Color usually gray.
Natural History: Occurs on sand or mud bottoms in depths of 40 to 1150 feet.
Size: Reaches 40 cm in diameter.
Range: Southeast Alaska to San Diego, California.

134. FAT SEA STAR

135. SPINY SEA STAR

136. SPINY SAND STAR

137. SAND STAR

138. RAINBOW SEA STAR ***Orthasterias koehleri***
(Koehler's straight star)

Identification: A colorful, five-armed star with long, slender, spiny arms. The color is usually a mottled red, purple and white.
Natural History: A common rocky reef sea star. They prey on abalones, squid, crabs and fish. Rainbow sea stars occur as shallow as the low intertidal and as deep as 2190 feet.
Size: Reaches 38 cm. in diameter.
Range: Alaska to southern California.

139. FRAGILE RAINBOW STAR ***Astrometis sertulifera***
(Garland-bearing star craft)

Identification: The five long, slender arms are covered with red-tipped spines. Color is usually green-brown, sometimes mottled yellow, orange or purple.
Natural History: Can be observed on rocky reefs in depths from the intertidal out to at least 60 feet, although, because of their protective coloration, they are often difficult to see. Feed on snails, clams and barnacles. These beautiful sea stars readily shed arms when handled.
Size: Maximum size is 25 cm. in diameter.
Range: Santa Barbara, California, to Baja California and the Gulf of California.

140. FISH-EATING SEA STAR ***Stylasterias forreri***
(Forrer's column star)

Identification: A brownish sea star with five long, slender arms covered with spines and pedicellaria.
Natural History: Fish-eating sea stars can be observed on rocky reefs usually in or near crevices in depths of 25 to 2190 feet. The brown-gray color and numerous pedicellaria on the arms allow these sea stars to capture and consume small fish which inadvertently land on their arms.
Size: May reach 50 cm. in diameter.
Range: Southern Alaska to San Diego, California.

141. FALSE OCHRE SEA STAR ***Evasterias troschellii***
(Troschell's true star)

Identification: A usually orange sea star with white spines on five long, slender arms, and a small disc. Other colors include gray, gray-green, brown or nearly red.
Natural History: False ochre sea stars feed on mussels, clams and barnacles that inhabit rocky reefs, pier pilings or other hard substrates, in depths ranging from the intertidal out to 230 feet.
Size: May reach 60 cm. in diameter.
Range: Pribiloff Islands to Carmel Bay, California.

138. RAINBOW SEA STAR

139. FRAGILE RAINBOW STAR

140. FISH-EATING SEA STAR

141. FALSE OCHRE STAR

142. **OCHRE STAR** ***Pisaster ochraceus***

(Ocher sea star)

Identification: Look for five relatively short, thick arms and a large disc area.
Natural History: The ochre star is the most common sea star of the rocky intertidal. They are also common members of the rocky subtidal community. They feed on mussels, barnacles, clams and small abalone and occur to depths of 80 feet.
Size: Reaches 35 cm. in diameter.
Range: Southern Alaska to Cedros Island, Baja California.

143. **GIANT-SPINED SEA STAR** ***Pisaster giganteus***

(Giant sea star)

Identification: A five-armed star with white spines surrounded at the base by a blue ring.
Natural History: Occurs in the rocky intertidal out to rocky reefs as deep as 100 feet. The giant-spined sea star preys on clams, snails and abalone. A very common member of the central and southern California reef community.
Size: Reaches 55 cm. in diameter.
Range: Vancouver Island to Cedros Island and Pablo Bay, Baja California.

144. **SHORT-SPINED SEA STAR** ***Pisaster brevispinus***

(Short-spined sea star)

Identification: A typical *Pisaster*-shaped, five-armed, star. It lacks long spines and is usually pink in color.
Natural History: Occurs on sand or mud as well as rocky bottoms from the intertidal out to depths of 300 feet or more. Like the other *Pisasters*, the short-spined sea star feeds on clams and other creatures of the soft-bottom community.
Size: Reaches 60 cm. in diameter.
Range: Alaska to La Jolla, California.

145. **SPINY BRITTLE STAR** ***Ophiothrix spiculata***

(Spiky snake-hair)

Identification: Each of the five arms has long, serrated spines. The disc also has spines. Color variable—blue, green, orange, red and yellow.
Natural History: Occurs on rocky reefs from the intertidal out to depths of 250 feet. Females carry eggs in sacs until the larvae leave. Brittle stars feed on bottom detritus and small animals, living or dead. They are also able to collect food suspended in the water column.
Size: Reaches 12 cm. in diameter.
Range: Central California to Central America.

142. OCHRE STAR

143. GIANT-SPINED STAR

144. SHORT-SPINED STAR

145. SPINY BRITTLE STAR

146. **SMOOTH BRITTLE STAR** ***Ophioplocus esmarki***

(Esmark's snake braid)

Identification: A brittle star with five short, flattened, relatively smooth arms. Color is usually gray to gray-brown.
Natural History: Occurs on sand or mud bottoms and also rocks from the intertidal to depths of 40 feet. Females brood eggs.
Size: Reaches about 10 cm. in diameter; maximum disc size is about 2 cm.
Range: Central California to Baja California.

147. **COMMON BASKET STAR** ***Gorgonocephalus caryi***

(Cary's gorgon-head)

Identification: A brittle star with a multitude of branching arms. This is the only basket star in the Northeastern Pacific.
Natural History: Occurs on rocky reefs and soft bottoms in depths of 50 to 3000 feet. Basket stars are nocturnal feeders—at least those living in shallow waters. They feed by extending arms into the water column and capturing small animals as they drift by. The food is then delivered to the mouth by the successful arm.
Size: Reaches 75 cm. in height, arms extended.
Range: Gulf of Alaska to California.

148. **CALIFORNIA SEA CUCUMBER** ***Parastichopus californicus***

(California rows-of-feet)

Identification: A large cucumber possessing pointed papillae. The tube feet are only on the underside of the animal. Color ranges from brown to reddish-brown.
Natural History: Common in central California on and around rocky reefs,and soft bottoms in depths ranging from intertidal to 100 feet. The California cucumber feeds on detritus by taking in large amounts of bottom sand or mud and extracting the nutrients as the material passes through the body. Cucumbers are a prized delicacy in some parts of the world.
Size: Reaches 45 cm. in length.
Range: Gulf of Alaska to San Diego, California.

149. **WARTY SEA CUCUMBER** ***Parastichopus parvimensis***

(Twig-like + rows-of-feet)

Identification: Lacks tube feet on upper surface. Possesses numerous small, black-tipped tubercles.
Natural History: Common on and around reefs in southern California and on rocky as well as mud or sand bottoms. Their depth occurrence ranges from the intertidal out to at least 100 feet. The warty sea cucumber's life history is similar to that of the California sea cucumber.
Size: Reaches 45 cm. in length.
Range: Carmel Bay, California, to Cedros Island, Baja California.

146. SMOOTH BRITTLE STAR

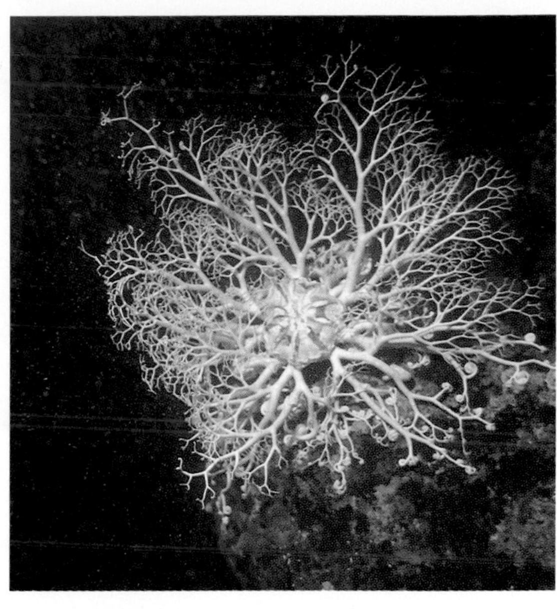

147. COMMON BASKET STAR

148. CALIFORNIA SEA CUCUMBER

149. WARTY SEA CUCUMBER

150. **ORANGE SEA CUCUMBER** ***Cucumaria miniata***

(Vermillion cucumber)

Identification: A large cucumber with multi-branched tentacles and five rows of tube feet. Color is usually orange, but can range from white to almost red.
Natural History: Occurs on rocky reefs in crevices from the intertidal out to 80 feet. In contrast to the two previously discussed sea cucumbers, the orange sea cucumber does not move about. The orange cucumber's feeding habits also differ in that they use their well-developed tentacles to capture food as it passes by. Once loaded with food, the tentacle is put into the mouth where the food is removed.
Size: Reaches 25 cm. in length.
Range: Gulf of Alaska to Avila Beach, California.

151. **WHITE SEA CUCUMBER** ***Eupentacta quinquesemita***

(Five true rays + five paths)

Identification: Look for the very long, non-retractable, branched tentacles. Color ranges from white to light orange.
Natural History: Occurs on rocky reefs from the intertidal out to depths of 40 to 50 feet. Usually found in crevices.
Size: Reaches 10 cm. in length.
Range: Sitka, Alaska, to Sacramento Reef, Baja California.

152. **SWEET POTATO** ***Molpadia arenicola***

(Sandy *Molpadia*—a woman's name)

Identification: A smooth, tough-skinned cucumber lacking tube feet. It is shaped like a sweet potato. The color is mottled bright yellow and brown.
Natural History: Usually found partially buried in sand in shallow water. They feed by passing large amounts of detritus through their gut where it is digested.
Size: Reaches 25 cm.
Range: Southern California to Baja California.

153. **SLIPPER SEA CUCUMBER** ***Psolus chitonoides***

(Penis [with a] coat-of-mail)

Identification: This flattened cucumber has distinct plates on top, and tube feet below. They are very difficult to dislodge.
Natural History: This sedentary cucumber can be found firmly attached to rocks, commonly in caves and crevices from the low intertidal out to depths of 50 feet. The slipper sea cucumbers utilize their well-developed tentacles to capture food.
Size: Reaches 12 cm. in length.
Range: Gulf of Alaska to Baja California.

150. ORANGE SEA CUCUMBER

151. WHITE SEA CUCUMBER

152. SWEET POTATO

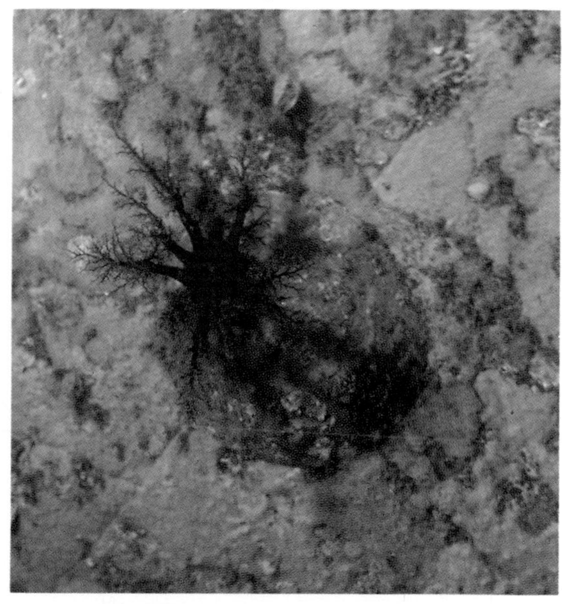

153. SLIPPER SEA CUCUMBER

PHYLUM CHORDATA

SUBPHYLUM UROCHORDATA

TUNICATES

154. **LOBED TUNICATE** ***Cystodytes lobata***

(Lobed pouch entrance)

Identification: This colonial tunicate is an encrusting form that can become quite thick. The upper surface is often arranged in convolutions or irregular ridges. The tunic is translucent and the zooids within give it a gray to pink color.
Natural History: *Cystodytes* is found in protected and exposed rocky areas from the low intertidal to 100 feet.
Size: Up to 3 cm. thick and 25 cm. diameter.
Range: Oregon to Baja California.

155. **ELEPHANT EAR TUNICATE** ***Polyclinum planum***

(Flat, very sloped)

Identification: This colonial tunicate is in the form of a single, thick, flattened ear-like lobe. The openings to the zooid chambers appear as minute craters on the surface. It is attached by a short, stout stalk (peduncle). The color is brown to orange.
Natural History: This is a fairly common species along the Big Sur coast and south of Pt. Conception. It attaches to rock in semi-protected open coastal areas from the low intertidal to about 80 feet.
Size: Maximum height about 15 cm.; width about 10 cm.
Range: Fort Bragg, California, to southern California.

156. **LIGHT-BULB TUNICATE** ***Clavelina huntsmani***

(Huntsman's little club)

Identification: A club-shaped, translucent tunic encloses this colonial animal. Inside can be seen the pink internal organs that resemble a filament in a light bulb.
Natural History: This tunicate usually occurs in clusters under rock ledges, on rock walls and often among the hold-fasts of kelp. It is apparently an annual species, reaching its peak during summer and degenerating through the fall when most of them disappear. Occurs from the intertidal to 100 feet.
Size: Length to about 2.5 cm.
Range: Vancouver Island, British Columbia, to Pt. Conception.

157. **SEA PEACH** ***Halocynthia aurantia***

(Orange Cynthis [a moon goddess] of the sea)

Identification: Although the shape is oval, this tunicate derives its common name from a resemblance to a peach in its smooth skin and orange coloration. The prominent siphons are located at its upper end.
Natural History: The Pacific sea peach, which has a similarly named Atlantic relative, is found attached to rocks along outer shores from the shallow subtidal to 100 feet. Like most tunicates, it is hermaphroditic, possessing both male and female reproductive organs in a single animal.
Size: Maximum height to about 15 cm.
Range: Alaska to British Columbia.

154. LOBED TUNICATE

155. ELEPHANT EAR TUNICATE

156. LIGHT-BULB TUNICATE

157. SEA PEACH

158. **BROAD-BASE TUNICATE** ***Cnemidocarpa finmarkiensis***
(Finmark's fruit [with] leggings)

Identification: The color of this smooth-skinned tunicate is bright orange to nearly red. The body is low and rounded and its short contractible siphons are tube-like. By stretching one's imagination a little, it almost looks like a "fruit with leggings".
Natural History: *Cnemidocarpa* is found on exposed rock surfaces that frequently are overgrown with smaller organisms or covered with dusty-looking debris—a marked contrast to its own smooth and clean appearance. Intertidal to 100 feet.
Size: Maximum diameter about 5 cm.
Range: Alaska to Pt. Conception.

159. **STALKED TUNICATE** ***Styela montereyensis***
(Monterey pillar)

Identification: This solitary animal has a long, thin stalk. Its surface is wrinkled and leathery and the most common color is an orangish-red. At the tip of the animal are its two siphons: one incurrent, the other excurrent.
Natural History: It occurs in clean water on hard surfaces such as rock and pier pilings and just sits there, subject to the vagaries of the currents, quietly feeding and growing. Growth is fairly rapid and its adult life-span is about 1-3 years. Intertidal to 100 feet.
Size: Maximum length about 25 cm.
Range: Alaska to San Diego.

160. **SPINY-HEADED TUNICATE** ***Boltenia villosa***
(Bolten's shaggy one)

Identification: The round bulbous "head" of this tunicate is attached to the substrate by a thin stalk that is often 75% of its total length. The body is covered with spinous projections and the siphons are not usually readily visible.
Natural History: *Boltenia* grows in habitat similar to *Styela* (#159), on rocks of the outer coast from the intertidal to 80 feet where there is generally clean water and strong currents. However, detritus still clings to the spinous exterior and usually obscures its orange-brown background coloration.
Size: Maximum length about 10 cm.
Range: Alaska to Pt. Conception.

161. **GLASSY TUNICATE** ***Ascidia paratropa***
(Nearly changed bladder)

Identification: The tunic of this animal is nearly translucent and covered with scattered, pointed papillae. The internal organs can be seen hazily through the tunic. The siphons are at the top of the animal.
Natural History: More common in the Pacific Northwest, from the shallow subtidal to about 150 feet, the glassy tunicate stands stiffly erect on rock or shell substrate. Within its branchial chamber lives several commensal invertebrate species, one of which is the hydroid, *Endocrypta huntsmanni*.
Size: Maximum height about 10 cm.
Range: Vancouver Island, British Columbia, to Monterey, California.

158. BROAD-BASE TUNICATE

159. STALKED TUNICATE

160. SPINY-HEADED TUNICATE

161. GLASSY TUNICATE

BIBLIOGRAPHY

Abbott, R. Tucker, 1968. *Seashells Of North America.* Golden Press, N.Y., 280 pp.

Furlong, Marjorie and Virginia Pill, 1970. *Starfish, Methods Of Preserving And Guides To Identification.* Ellison, Ind., Edmonds, Wash.

Johnson, Myrtle E., and Harry J. Snook, 1927. *Seashore Animals Of The Pacific Coast.* Dover Pub., N.Y., 659 pp.

Keen, A. Myra and Eugene Coan, 1974. *Marine Molluscan Genera Of Western North America,* 2nd edition. Stanford University Press, Calif., 208 pp.

Kozloff, Eugene N., 1973. *Seashore Life Of Puget Sound, The Strait Of Georgia And The San Juan Archipelago.* University of Washington Press, Seattle, Wash. 282 pp.

Mc Lean, James H., 1969. *Marine Shells of Southern California.* Los Angeles County Museum of Natural History, Exposition Park, Los Angeles, California. Science Series 24, Zoology No. 11, 104 pp.

Morris, Percy A., 1966. *A Field Guide To Shells Of The Pacific Coast And Hawaii.* Houghton Mifflin Co., Boston, Mass. 297 pp.

Morris, R. H., D. P. Abbott, and E. C. Haderlie, 1980. *Intertidal Invertebrates of California.* Stanford University Press, Stanford, California. 690 pp.

North, Wheeler, J., 1976. *Underwater California.* University of California Press, Berkeley, 176 pp.

Ricketts, Edward F. and Jack Calvin, 1968. *Between Pacific Tides.* 4th Ed. Revised by J. W. Hedgpeth. Stanford University Press, 614 pp.

Smith, Ralph I. and James T. Carlton (editors), 1975. *Light's Manual:* Intertidal Invertebrates of the Central California Coast. Third ed. University of California Press, Berkeley, 716 pp.

INDEX TO COMMON NAMES

INDEX TO SCIENTIFIC NAMES

R

S

T

U

V

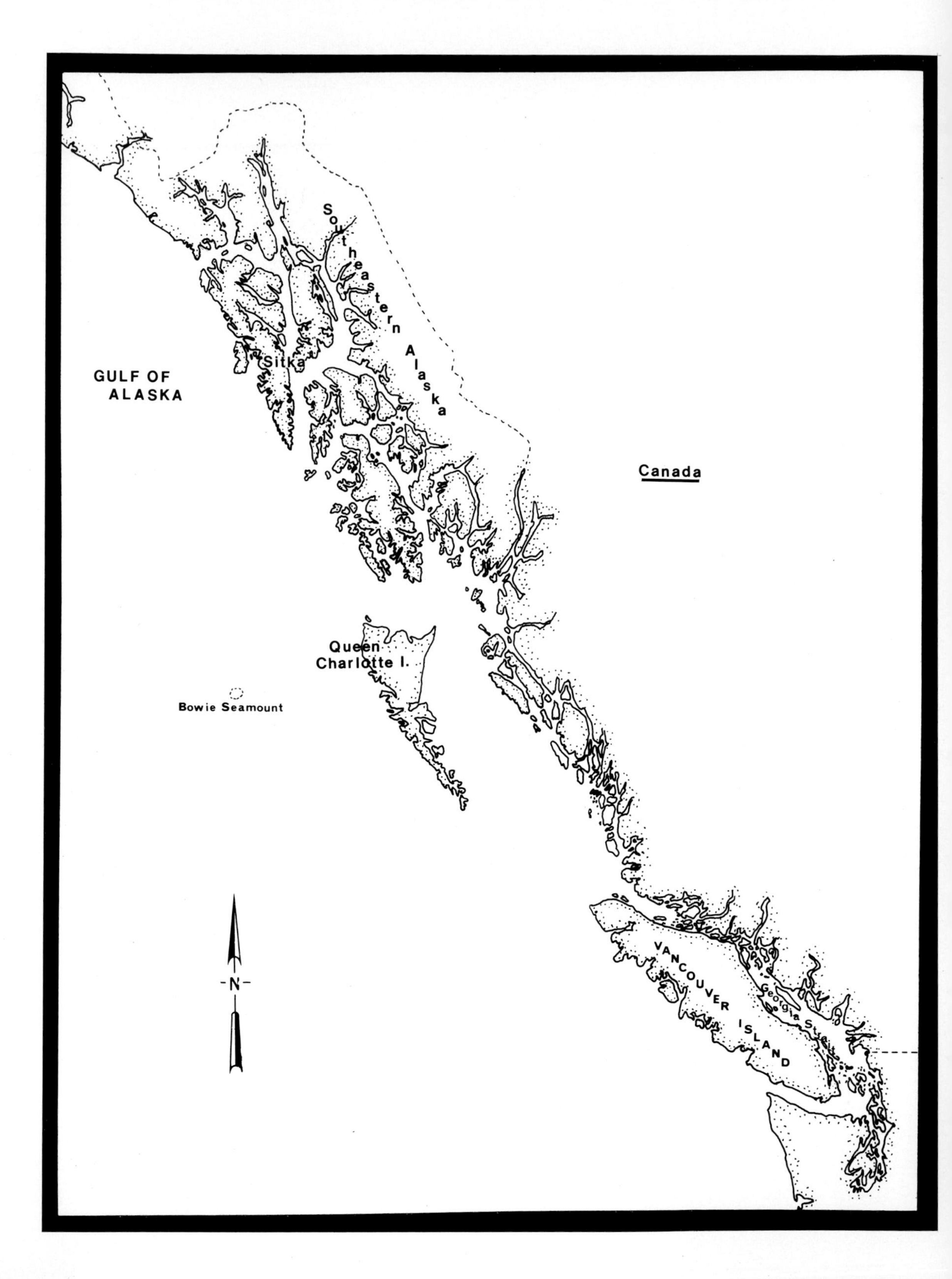

GULF OF ALASKA
Southeastern Alaska
Sitka
Canada
Queen Charlotte I.
Bowie Seamount
VANCOUVER ISLAND
Georgia Strait
N

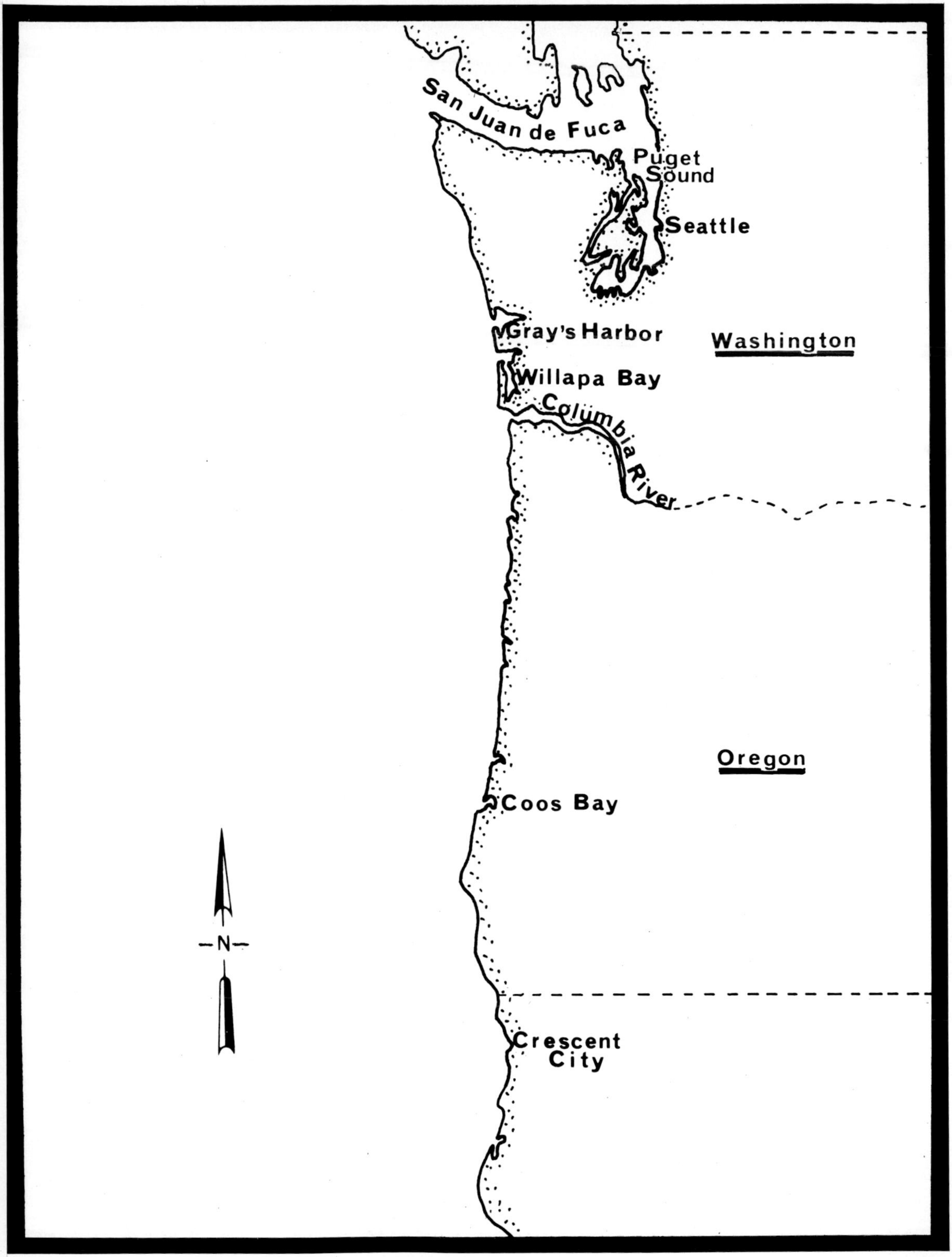

San Juan de Fuca

Puget Sound

Seattle

Gray's Harbor

Washington

Willapa Bay

Columbia River

Oregon

Coos Bay

Crescent City

N

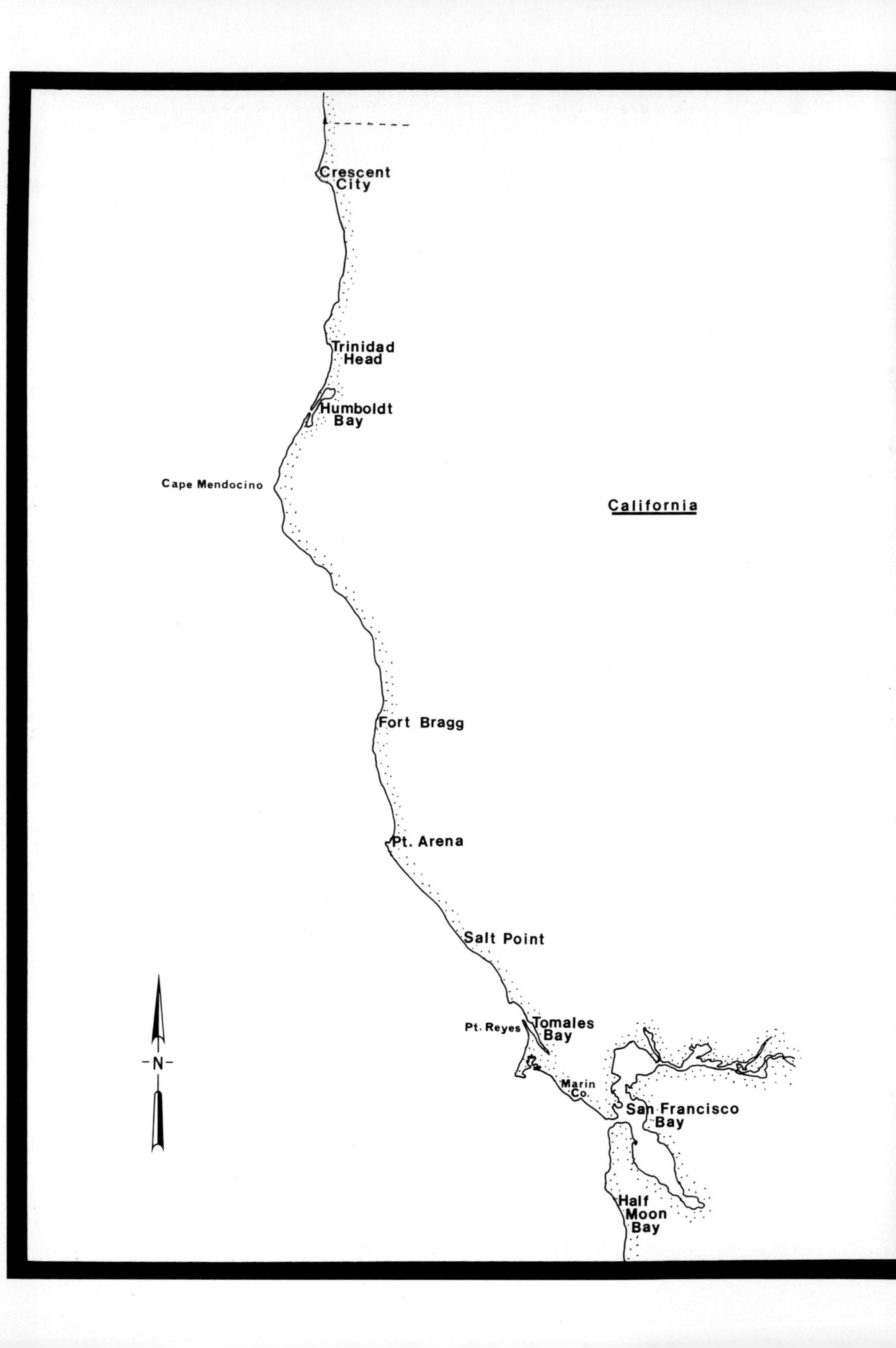

Crescent City
Trinidad Head
Humboldt Bay
Cape Mendocino
California
Fort Bragg
Pt. Arena
Salt Point
Tomales Bay
Pt. Reyes
N
Marin Co.
San Francisco Bay
Half Moon Bay

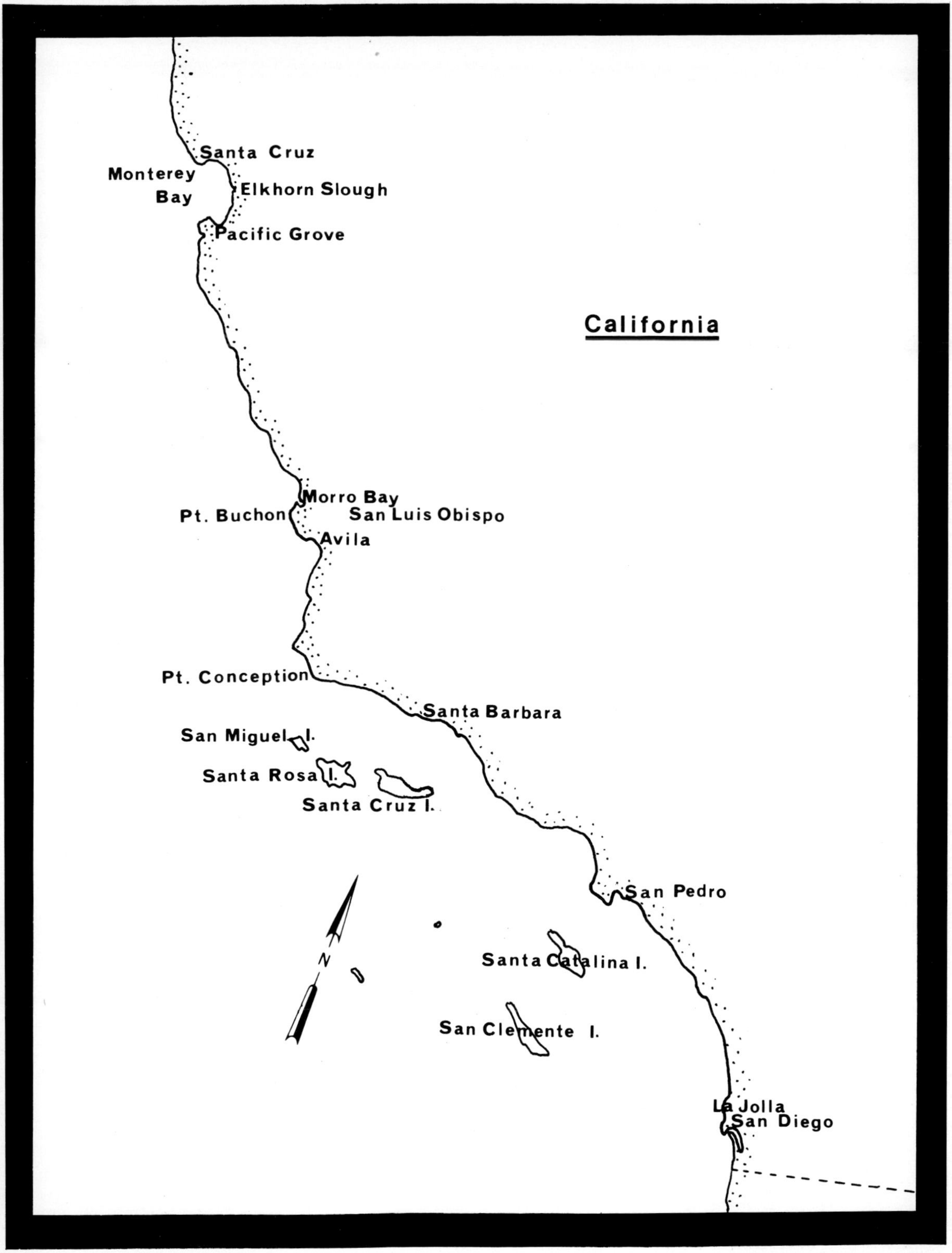

California

Santa Cruz
Monterey Bay
Elkhorn Slough
Pacific Grove
Morro Bay
Pt. Buchon
San Luis Obispo
Avila
Pt. Conception
Santa Barbara
San Miguel I.
Santa Rosa I.
Santa Cruz I.
San Pedro
N
Santa Catalina I.
San Clemente I.
La Jolla
San Diego

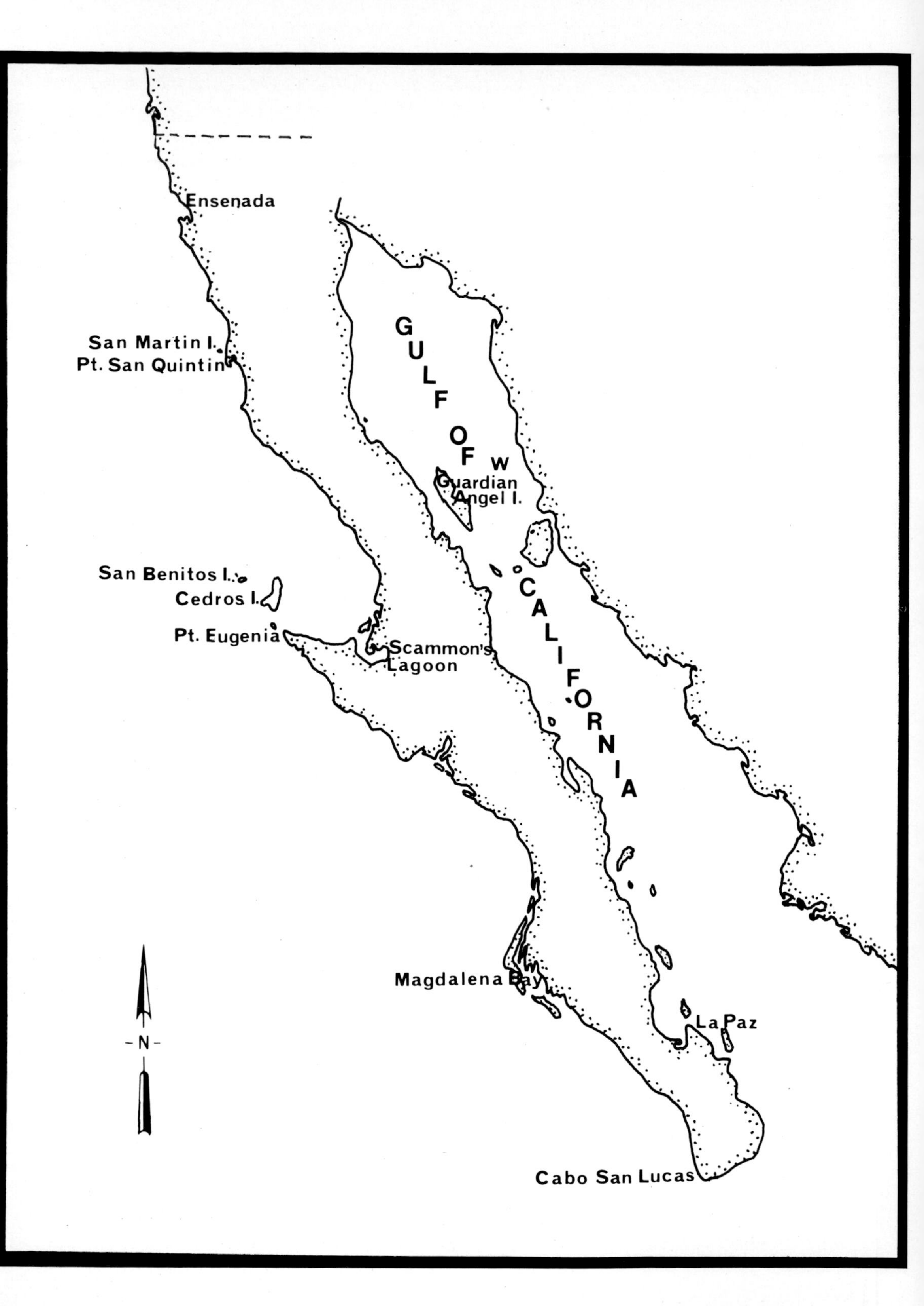
Ensenada
San Martin I.
Pt. San Quintin
GULF OF CALIFORNIA
Guardian Angel I.
San Benitos I.
Cedros I.
Pt. Eugenia
Scammon's Lagoon
Magdalena Bay
La Paz
Cabo San Lucas
N